Neuropsychiatric Dysfunction
in Multiple Sclerosis

Ugo Nocentini • Carlo Caltagirone
Gioacchino Tedeschi
Editors

Neuropsychiatric Dysfunction in Multiple Sclerosis

 Springer

Editors
Ugo Nocentini
Dipartimento di Neuroscienze
Università degli Studi di Roma
"Tor Vergata"
Rome
Italy

and

IRCCS
Fondazione Santa Lucia
Rome
Italy

Carlo Caltagirone
Dipartimento di Neuroscienze
Università degli Studi di Roma
"Tor Vergata"
Rome
Italy

and

IRCCS
Fondazione Santa Lucia
Rome
Italy

Gioacchino Tedeschi
Istituto di Scienze Neurologiche
Seconda Università di Napoli
Naples
Italy

The contents of this book are based on the Italian edition published under the title *I disturbi neuropsichiatrici nella sclerosi multipla*. U. Nocentini, C. Caltagirone, G. Tedeschi (eds.)
© Springer-Verlag Italia 2011

ISBN 978-88-470-2675-9   ISBN 978-88-470-2676-6 (eBook)
DOI 10.1007/978-88-470-2676-6
Springer Milan Heidelberg New York Dordrecht London

Library of Congress Control Number: 2012949407

Printed on acid-free paper

Springer is part of Springer Science+Business Media (www.springer.com)

Foreword

Multiple Sclerosis (MS) is a disease of the nervous system which has profound effects on everyday life of not only the person who is diagnosed, but her/his family and close friends. MS is among the most common cause of neurological disability among young adults, and affects about 2.5 million people worldwide. In the United States, a new person is diagnosed with MS about every hour. Fortunately, we are in the midst of an intellectual explosion in understanding the underlying neurobiology of this disease. This is largely due to the rapid and significant advances in neuroimaging over the last decade, in addition to considerable new knowledge concerning the basic neuropathology of MS. For example, while classically viewed as a white matter disease involving myelin in the central nervous system (CNS), we now know that MS also involves CNS gray matter, and that pathology is also found in the cerebral cortex where it was previously delimited to "subcortical" regions. These advances in neurobiology have led to new and improving treatments for persons with MS. In fact, work over the past two decades is a testament to a translational approach to clinical science, whereby basic knowledge results in new treatment in clinical practice. These treatments almost exclusively involve slowing the progression of the disease. Although modest in effectiveness, it represents a significant leap forward in our approach to the disease and holds promise for new and more effective therapies in the near future. On the other hand, symptomatic treatment in MS seems to have lagged behind, especially involving the neuropsychiatric consequences of MS. For example, there are no approved pharmacological treatments for cognitive impairment, and behavioral approaches have been mixed, limited, and marred by significant methodological limitations, which haunt any potential applicable benefits. One very significant void in the work on MS to date is the lack of evidence-based treatment for progressive forms of the disease. Therefore, despite the significant advances in our knowledge of the neurobiology of MS, little transfer of this knowledge toward treatment of progressive MS has resulted, and it is an area that requires immediate attention.

Despite the significant advances made over the last decade in understanding the pathophysiology of MS through work in neuroimaging, we also have become increasingly aware about the lack of a one to one relationship between MS symptoms and MS pathology. For instance, the majority of studies involving cognition show that pathology observed via neuroimaging account for less than 50% of the variance, and oftentimes considerably less. There are clearly individual differences that account for the less-than-desired relationship between neuroimaging parameters and behavior, which are both genetic and environmentally determined. We are only beginning to understand this relationship. For instance, it is only recently, that we have found that lifetime intellectual enrichment of the MS patient has a profound impact on the expression of cognitive symptoms, a concept referred to as "cognitive reserve." That is, persons with a high degree of lifetime intellectually challenging experiences are significantly less likely to display cognitive impairments relative to those with less fortunate intellectual stimulation, even when both have the same degree of brain atrophy on MRI. Clearly, the impact of environmental and genetic factors that mediate the expression of the disease itself, as well as our ability to assess these symptoms through existing and future instruments, is a critical area of future research.

Even with the recent advances that have come with neuroimaging, relatively little is known about the neuropsychiatric consequences of the disease. There is a paucity of comprehensive publications integrating the vast spectrum of the disease's behavioral consequences, ranging from cognitive, behavioral, personality, and psychopathology, that exist. This new volume, *Neuropsychiatric Dysfunctions in Multiple Sclerosis*, edited by Nocentini, Caltagirone, and Tedeschi, fills this void. Not only does it provide detailed and current knowledge on numerous aspects of the disease itself (e.g., epidemiology, pathophysiology, diagnostic and clinical manifestations, therapy, and rehabilitation) but it provides a detailed overview of the neuropsychiatric consequences of MS. This is a very unique contribution to the MS discussion. Clinicians are often faced with very complicated behavioral and psychiatric complications in an MS patient, and they have limited resources to reference other than a time-consuming literature review that is often selective and lacks integration. This new book provides an up-to-date and comprehensive presentation, which will be extremely helpful for the clinician. Its focus on neuropsychiatric manifestation of the disease provides the MS student a single volume in which comprehensive knowledge on the topic is contained and the MS scholar the appropriate tools necessary to envision future directions in this critically important area of MS research.

John DeLuca, Ph.D., ABPP
Vice President for Research
Professor of Physical Medicine and Rehabilitation
and of Neurology and Neurosciences
UMDNJ-New Jersey Medical School
Newark, USA

Acknowledgments

First, we would like to sincerely thank the patients we have encountered during our professional activity. As a matter of fact, their lives are intertwined with ours. Even if at times their suffering affected us quite much, we hope we have always shown them our humanity as well as our professional skills. We are grateful to the patients and their families because we learned from them much more than from books, scientific journals, and conferences.

Still, with regard to what we have learned, we would like to thank our teachers for imparting their knowledge to us and providing an example of dedication to the values of life. In any case, we will all be gratified if in the future we are esteemed in the same way as we consider those who have supported us during the years of our training.

We also wish to thank our families for their patience when we had to sacrifice moments of life we wished we could share with them.

Furthermore, special thanks goes to the institutions where we work for providing material and moral support for our clinical and research work.

In addition, we would like to thank our readers in advance for choosing this book; we hope that they will find it useful and enlightening.

Last but not least, we would like to thank the publisher, Springer, for supporting our initiative. Above all we would like to thank those people working in Springer-Verlag Italia for helping us complete this work. In particular, we thank Catherine Mazars, Juliette Kleemann, Roberto Garbero, and those who work in the administration offices and typography. We believe that being able to speak of a good result is only due to the efforts of all those who contributed to making this book available to the readers.

Contents

Part I   Multiple Sclerosis: General Clinical Aspects

1   Introduction   3
Silvia Romano, Carlo Caltagirone, and Ugo Nocentini

2   Epidemiology   7
Silvia Romano, Carlo Caltagirone, and Ugo Nocentini

3   Clinical Presentation   11
Silvia Romano, Carlo Caltagirone, and Ugo Nocentini

4   Etiopathogenesis   21
Silvia Romano, Carlo Caltagirone, and Ugo Nocentini

5   Neuropathology   27
Silvia Romano, Carlo Caltagirone, and Ugo Nocentini

6   The Diagnosis of Multiple Sclerosis   31
Simona Bonavita and Gioacchino Tedeschi

7   Assessment Instruments   37
Silvia Romano, Carlo Caltagirone, and Ugo Nocentini

8   Neuroimaging in Multiple Sclerosis   43
Gioacchino Tedeschi, Renato Docimo, Alvino Bisecco,
and Antonio Gallo

9   Therapy of Multiple Sclerosis   65
Alessandro d'Ambrosio and Simona Bonavita

10   Rehabilitation in Multiple Sclerosis   77
Ugo Nocentini and Carlo Caltagirone

Part II Psychiatric Disturbances in Multiple Sclerosis

11 **Depression and Anxiety** 85
Alberto Siracusano, Cinzia Niolu, Lucia Sacchetti, and Michele Ribolsi

12 **Bipolar Disorder and Mania** 99
Alberto Siracusano, Cinzia Niolu, Michele Ribolsi, and Lucia Sacchetti

13 **Mood Dysfunctions in MS and Neuroimaging** 107
Antonio Gallo, Rosaria Sacco, and Gioacchino Tedeschi

14 **Psychosis** 113
Patrizia Montella, Manuela de Stefano, Daniela Buonanno,
and Gioacchino Tedeschi

15 **Euphoria, Pathological Laughing and Crying** 121
Silvia Romano and Ugo Nocentini

16 **Emotions and Multiple Sclerosis** 127
Ugo Nocentini

Part III Cognitive Dysfunctions and Multiple Sclerosis

17 **Cognitive Dysfunctions in Multiple Sclerosis** 133
Ugo Nocentini, Silvia Romano, and Carlo Caltagirone

18 **Conclusions** 155
Ugo Nocentini, Gioacchino Tedeschi, and Carlo Caltagirone

Index 159

Part I

Multiple Sclerosis: General Clinical Aspects

Introduction

1

Silvia Romano, Carlo Caltagirone, and Ugo Nocentini

Multiple sclerosis (MS) is a chronic inflammatory, demyelinating and degenerative disease of the central nervous system (CNS) with a highly variable course. The exact cause of inflammation remains unclear, but an autoimmune response directed against CNS antigens is suspected. MS is the second most common neurologic cause of disability of young adults, after head trauma [1] (about 2.5 million people affected worldwide, 350,000 in Europe alone).

During the course of the disease a wide range of functional impairments and disabilities may develop, leading to a significant impact on the quality of life of patients, their family members and on employment (unemployment rates are higher for people with MS) [2]. According to the World Health Organization (WHO), MS is one of the most expensive diseases, with an annual social cost of 1 billion 600 million euros in Italy alone and an average annual cost per person of about 32,000 euros.

MS is classified in the group of CNS demyelinating diseases. This is a heterogeneous group of neurologic disorders characterized by a relatively selective damage of the CNS myelin. Demyelinating diseases may be classified as primary, characterized by direct damage of myelin, or secondary, in which the damage to the myelin results from neuronal or axonal injury. The spectrum of primary forms includes MS, acute disseminated encephalomyelitis (ADEM), optic neuritis and a group of diseases considered as particular variants of MS (Balo's concentric sclerosis, Schilder's diffuse cerebral sclerosis, Marburg's disease, neuromyelitis optica or Devic's disease).

S. Romano
Centre for Experimental Neurological Therapies (CENTERS), Neurology Unit, S. Andrea Hospital, University of Rome "La Sapienza", Rome, Italy
e-mail: silvia.romano@uniroma1.it

C. Caltagirone • U. Nocentini (✉)
Dipartimento di Neuroscienze, Università degli Studi di Roma "Tor Vergata", Rome, Italy
e-mail: c.caltagirone@hsantalucia.it; u.nocentini@hsantalucia.it

U. Nocentini et al. (eds.), *Neuropsychiatric Dysfunction in Multiple Sclerosis*,
DOI 10.1007/978-88-470-2676-6_1, © Springer-Verlag Italia 2012

A precise clinical description of MS was given for the first time by Jean Cruveilhier in 1835 [3], but it was not until some years later (1868) that the first scientific definition of clinical and neuro-pathological characteristics was provided when, in three lectures on disseminated sclerosis (lectures V, VI and VII) at the Hôpital de la Salpêtrière, Charcot clearly delineated the main features of the disease. He identified three distinct forms (spinal, cephalic and bulbar), characterized by the triad of symptoms intention tremor, nystagmus and scanning speech (now known as Charcot's triad) and described the neuro-pathological damage with loss of myelin, glial scar formation and consequent axonal injury [4]. However, the first description of a case suggestive of MS dates from the fourteenth century, in the biography of St Ludwina of Schiedam. The extremely detailed documentation of the saint's life preserved in the Vatican Archives records that St Ludwina's disease began at the age of 16 years with remitting motor disorders followed by a slowly progressive course.

All autoimmune diseases such as MS are thought to be caused by dysregulation of the immune system, with the formation of immune cells specifically activated against CNS components; these cells are able to adhere to vessel walls, extravasate and migrate into nervous tissue, where they attack the myelin sheaths that preserve nerve fibers.

It is well known that the main feature of this disease is an inflammatory process resulting in the loss of the myelin sheath and subsequent axonal degeneration, with onset of symptoms/signs of focal CNS injury. The demyelinating process mainly involves white-matter long tracts and periventricular white matter, the optic nerve, the spinal cord, the brainstem and the cerebellum. The symptoms of the acute phase (first episode and relapses) tend initially to regress spontaneously, but over time they cause permanent neurological deficits. The acute clinical episodes are characterized by focal inflammatory lesions, identified as contrast-enhancing lesions on magnetic resonance imaging (MRI); in more advanced stages, when the neurologic deficits remain clinically stable, MRI shows signs of neurodegeneration characterized by diffuse atrophy in addition to the hyperintense white matter lesions. Although MS is considered a demyelinating disease, in recent years neuro-pathological and neuro-imaging studies have demonstrated the widespread involvement of not only the white matter but also the gray matter of the CNS [5, 6].

There is currently no cure for this disease, but the introduction of drugs that can change its course (interferons and glatiramer acetate) has resulted in a reduction of disease activity on clinical (number of clinical relapses) and MRI measures and a delay in disability progression [7]. In recent years, moreover, new drugs such as monoclonal antibodies (e.g., natalizumab) or selective, orally administered, immune-suppressors (e.g., fingolimod) have become available, and several phase III trials with new drugs are in progress.

References

1. Adams RD, Victor M, Ropper AH (1997) Principles of neurology. McGraw-Hill, New York
2. Benito-Leon J, Morales JM, Rivera-Navarro J, Mitchell A (2003) A review about the impact of multiple sclerosis on health-related quality of life. Disabil Rehabil 25:1291–1303
3. Cruveilhier J (1835) Anatomie pathologique du corps humain; descriptions avec figures lithographiees et coloriees; des diverses alterations morbides dont le corps humain est susceptible. JB Bailliere, Paris
4. Charcot JM (1877) Lectures on the diseases of the nervous system delivered at the Salpêtrière. New Sydenham Society, London
5. Geurts JJ, Barkhof F (2008) Grey matter pathology in multiple sclerosis. Lancet Neurol 7:841–851
6. Chard D, Miller D (2009) Grey matter pathology in clinically early multiple sclerosis: evidence from magnetic resonance imaging. J Neurol Sci 282:5–11
7. Javed A, Reder AT (2006) Therapeutic role of beta-interferons in multiple sclerosis. Pharmacol Ther 110:35–56

Epidemiology

2

Silvia Romano, Carlo Caltagirone, and Ugo Nocentini

MS has been described in various populations and geographical regions with different frequencies.

Generally, the most commonly used frequencies in epidemiological studies of diseases are prevalence and incidence.

As described in the next subsection (etiopathogenesis), regional differences in MS prevalence and incidence have contributed to formulating a hypothesis on its pathogenesis.

Prevalence is traditionally defined as a measurement of the proportion of "events" in a population at a given point of time. "Event" is defined as the occurrence of any phenomenon that can be discretely characterized, for example: infection, presence of antibodies, pregnancy. In epidemiology, diseases or infections are the most commonly used events, and therefore prevalence can be defined as the proportion of people in a population (of a state, region, province) that, at any given time, are affected by the disease. Incidence, on the other hand, measures the number of new cases of a disease occurring during a given period (for example, in a month or a year) and identifies the risk (i.e. probability) of developing the disease in the considered population. Since incidence indicates a change in a quantity (new people affected) compared with the change in another quantity (time), it is considered to be a dynamic measure.

Epidemiological studies, as well as other studies, are made difficult by the particular nature of MS. In fact, epidemiological researches aimed at formulating and testing etiopathogenetic hypotheses demand levels of diagnostic accuracy

S. Romano
Centre for Experimental Neurological Therapies (CENTERS), Neurology Unit, S. Andrea Hospital, University of Rome "La Sapienza", Rome, Italy
e-mail: silvia.romano@uniroma1.it

C. Caltagirone • U. Nocentini (✉)
Dipartimento di Neuroscienze, Università degli Studi di Roma "Tor Vergata", Rome, Italy
e-mail: c.caltagirone@hsantalucia.it; u.nocentini@hsantalucia.it

U. Nocentini et al. (eds.), *Neuropsychiatric Dysfunction in Multiple Sclerosis*,
DOI 10.1007/978-88-470-2676-6_2, © Springer-Verlag Italia 2012

which overlap in different geographical regions so that diagnostic criteria shared by the scientific community and applicable in different clinical settings are required.

In the case of MS, in the past but in the present too, the accuracy and application of well- defined and uniform criteria (such as the use of MRI criteria for MS diagnosis) cannot be taken for granted. In considering the accuracy of MS epidemiological studies, there is also the difficulty of identifying a control group of suitable size and characteristics. While presenting a higher incidence in a specific age range, MS may occur from childhood until the seventh decade; in particular, the onset of the disease often remains unrecognized for many years.

Another aspect to be considered is the role that genetic and racial factors play in susceptibility to the disease; these factors, although not yet clearly identified, are undoubtedly relevant variables.

Despite the above-mentioned difficulties, several epidemiological studies have been published [see 1–3] since the first prevalence surveys conducted from 1926 to 1929 [4–6].

The worldwide prevalence is estimated to be about 2.5 million people, 400,000 of whom live in the USA and 350,000 in Europe. Some studies have measured the death rate due to MS in a particular nation or geographic area, or in different geographic areas within the same nation, analyzing this event at the same time or different times. Other studies have aimed to identify prevalence or incidence rates as discussed above, in this case too taking account of the geographical and temporal parameters already mentioned. All these studies obviously consider the differences related to sex and race, but also other aspects related to social, economic and cultural factors. The most important information emerging from these studies is: mortality rates due to MS appear to be higher in temperate regions than in the tropics and subtropics, higher in Europe and North America than in Africa, South America, Asia and Mediterranean regions, higher in women than in men, in Caucasians than in non-Caucasians while, at least in the United States, the different rates between urban and rural areas do not seem to be significant.

Prevalence studies have shown a geographic distribution characterized by three distinct areas of disease frequency related to latitude. It is generally accepted that the prevalence of MS tends to increase with increasing distance from the equator. High-risk areas including Northern Europe, Southern Canada and Northern USA; medium-risk areas including southern Europe, the southern United States and Australia; and finally, low-risk areas including Asia, Alaska and Africa have been described. The current prevalence rates (50–80 cases/100,000 population) put Italy among the countries with the highest risk of the disease. In Italy MS incidence is approximately 3–8 cases/100,000 population per year, with significant regional variations and peaks in Sicily and Sardinia, reaching 45 cases/100,000 population. However, this hypothesis, based on latitude gradient, is not applicable in the most methodologically adequate studies designed to make reliable comparisons between incidence rates and prevalence rates [2]. A meta-analysis of epidemiological studies published between 1980 and 1998, which standardized rates by sex and age applied to the European and the world population, found no correlation between MS frequency and latitude [7]. Small clusters with high prevalence were identified in northwestern Sardinia [8, 9], Sicily [10, 11] and in the north of Croatia [12].

In the United Kingdom and North America, the risk of MS developing in immigrants from the Far East remains low, while the risk increases in the second generation of immigrants from India [13]. These differences between immigrants and their descendants could reflect very early exposure to a possible environmental factor. Migration studies have led the way for research on possible environmental factors, suggesting a role of sun exposure in MS pathogenesis [14].

Studies carried out on populations from South Africa, Israel, Hawaii and immigrants in Britain correlate the risk of developing the disease with the place of residence during childhood [15–18].

People who migrate before age 15 to an area with a different risk from that of their original homeland, acquire the risk of their new homeland [16]. However, in Australia, an analysis performed on an extremely homogeneous population showed no effect related to migration age (cut-off age of 15 years), suggesting that the exposure risk covers a wider age range than initially hypothesized [18].

These data support the fundamental role of environmental factors in the pathogenesis of the disease, suggesting the hypothesis that MS is acquired long before its clinical onset. However, migration studies also have limitations due to estimates made on rates originating from samples which are numerically not representative and are not able to identify the exact moment when the exposure to environmental factor or factors occurred (cumulative effect).

MS usually begins at an age between 25.3 and 31.8 years, with an incidence peak at around 29.2 years [19]. However, cases of pediatric MS (2.7 %) and cases with onset after 60 years are also described [20, 21].

The average age of onset tends to be lower in the eastern Mediterranean (26.9) and to increase in Europe (29.2), Africa (29.3), America (29.4), and South-East Asia (29.5) as far as the Western Pacific areas, where the highest values are reported (33.3). It is generally accepted that low-income countries have an average age of onset of 28.9 years, while the average age of onset in high-income countries is 29.5 [19].

MS mostly affects the female sex, with a male/female ratio of 0.5 (range 0.40–0.67); the ratio is lower in Europe (0.6) and higher in Africa (0.33) and in areas of the Western Pacific (0.31). There are no differences between the social classes [19]. The ratio is higher for onset around or after puberty (0.2–0.4), but it is inverted if only very early onset MS cases are considered [22]. The female/male ratio seems to increase over time: a Canadian longitudinal study found that MS in women has approximately tripled over the past 60 years and that the female to male sex ratio now exceeds three women with MS for every one man (3.2:1) [23]

References

1. Sadovnick AD, Ebers GC (1993) Epidemiology of multiple sclerosis: a critical overview. Can J Neuro Sci 20:17–29
2. Rosati G (1994) Descriptive epidemiology of multiple sclerosis in the 1980s: a critical overview. Ann Neurol 36:S164–S174
3. Kurtzke JF, Page WF (1997) Epidemiology of multiple sclerosis in US veterans: VII. Risk factors for MS Neurology 48:204–213
4. Bing R, Reese H (1926) Die Multiple Sklerose in der Nordwestschweiz. Schweiz Med Wschr 56:30
5. Ackermann A (1931) Die Multiple Sclerose in der Schweiz: Enquete von 1918–22. Schweiz Med Wschr 61:1245–1250
6. Allison RT (1931) Disseminated sclerosis in North Wales. Brain 53:391–430
7. Zivadinov R, Iona L, Monti-Bragadin L et al (2003) The use of standardized incidence and prevalence rates in epidemiological studies on multiple sclerosis. Neuroepidemiology 22:65–74
8. Sotgiu S, Pugliatti M, Sanna A et al (2002) Multiple sclerosis complexity in selected populations: the challenge of Sardinia, insular Italy. Eur J Neurol 9:329–341
9. Marrosu M, Lai M, Cocco E et al (2002) Genetic factors and the founder effect explain familial MS in Sardinia. Neurology 58:283–288
10. Grimaldi L, Salemi G, Grimaldi G et al (2001) High incidence and increasing prevalence of MS in Enna (Sicily), southern Italy. Neurology 57:1891–1893
11. Nicoletti A, Lo Fermo S, Reggio E et al (2005) A possible spatial and temporal cluster of multiple sclerosis in the town of Linguaglossa, Sicily. J Neurol 252:921–925
12. Materljan E, Sepcic J (2002) Epidemiology of multiple sclerosis in Croatia. Clin Neurol Neurosurg 104:192–198
13. Elian M, Nightingale S, Dean G (1990) Multiple sclerosis among United Kingdom-born children of immigrants from the Indian subcontinent, Africa and the West Indies. J Neurol Neurosurg Psychiatry 53:906–911
14. Acheson ED, Bachrach CA, Wright FM (1960) Some comments on the relationship of the distribution of multiple sclerosis to latitude, solar radiation and other variables. Acta Neurol Scand 35:132–147
15. Alter M, Halpern L, Kurland LT, Bornstein B, Leibowitz U, Silberstein J (1962) Multiple sclerosis in Israel. Prevalence among immigrants and native inhabitants. Arch Neurol 7:253–263
16. Dean G, Kurtzke JF (1971) On the risk of multiple sclerosis according to age at immigration to South Africa. Br Med J 3:725–729
17. Detels R, Visscher BR, Malmgren RM, Coulson AH, Lucia MV, Dudley JP (1977) Evidence for lower susceptibility to multiple sclerosis in Japanese-Americans. Am J Epidemiol 105:303–310
18. Hammond SR, English DR, McLeod JG (2000) The age-range of risk of developing multiple sclerosis: evidence from a migrant population in Australia. Brain 123:968–974
19. WHO. Atlas multiple sclerosis resources in the world 2008. WHO Library Cataloguing-in-Publication Data.
20. Noseworthy JH, Lucchinetti C, Rodriguez M, Weinschenker BG (2000) Multiple sclerosis. N Eng J Med 343:938–952
21. Compston A, Coles A (2002) Multiple sclerosis. Lancet 359:1221–1231
22. Ghezzi A, Pozzilli C, Liguori M et al (2002) Prospective study of multiple sclerosis with early onset. Mult Scler 8:115–118
23. Orton SM, Herrera BM, Yee IM et al (2006) Sex ratio of multiple sclerosis in Canada: a longitudinal study. Lancet Neurol 5:932–936

Clinical Presentation

3

Silvia Romano, Carlo Caltagirone, and Ugo Nocentini

The clinical characteristics of MS, both in terms of onset of disease (age, symptoms, clinical presentation) and in terms of short- and long-term evolution (course, relapse rate, speed and rate of progression) are highly variable across patients. There are many disease classifications that evaluate typical features of MS, such as course, clinical symptoms and signs, and location of lesions.

The best-known classification of MS clinical subtypes or forms has been developed based on the temporal profile. MS is usually classified in four subtypes [1]:

- Relapsing-remitting MS (RR) is the most common form of the disease, about 50% of cases, and is characterized by relapses followed by full or partial recovery of impaired functions, with a stable course and lack of progression during periods between relapses. A relapse or exacerbation is defined as an acute or sub-acute development of new symptoms, or objective signs of neurological dysfunctions, or the reappearance of preexisting symptoms with a duration greater than 24 h. If the symptoms occur within 30 days from the previous relapse, they are considered to belong to a single episode. Moreover, symptoms that occur during a febrile episode or in combination with other pathological conditions are also not defined as a relapse. In contrast, neurological deficits, even of short duration, occurring several times a day and for several days, have to be considered as a relapse. The relapse rate is extremely variable, as is the

S. Romano
Centre for Experimental Neurological Therapies (CENTERS), Neurology Unit, S. Andrea Hospital, University of Rome "La Sapienza", Rome, Italy
e-mail: silvia.romano@uniroma1.it

C. Caltagirone • U. Nocentini (✉)
Dipartimento di Neuroscienze, Università degli Studi di Roma "Tor Vergata", Rome, Italy
e-mail: c.caltagirone@hsantalucia.it; u.nocentini@hsantalucia.it

U. Nocentini et al. (eds.), *Neuropsychiatric Dysfunction in Multiple Sclerosis*,
DOI 10.1007/978-88-470-2676-6_3, © Springer-Verlag Italia 2012

outcome; relapses are generally more frequent during the first years and tend to decrease during the course of the disease;

- Secondary progressive MS (SP) is a form of the disease that occurs after the RR form (after a more or less long period), characterized by a continuing progression of neurological deficits, which may include relapses or stable phases and, more rarely, improvements. Both the time and way of transition from RR to SP subtype and progression speed are highly variable across patients;
- Primary progressive MS (PP) is characterized by progression of the disease from the onset, sometimes with stable periods or mild and temporary improvements;
- Progressive-relapsing MS (PR): this form is also characterized by a progressive course from the onset, but in this subtype clear acute relapse followed by a more or less complete recovery of impairments usually occurs.

There are other forms of MS, classified on the basis of severity of the disease and not yet well characterized: the malignant and benign forms. Malignant MS is defined as a disease with a rapidly progressive course leading to a severe disability in many functional systems or death in a relatively short period of time after disease onset. In contrast, benign MS is a form of the disease characterized by the absence of disability or by mild disability (EDSS $\leq$ 3.5) after a period of 15 years from disease onset [2]. However, recent studies have shown that even the so-called benign forms show significant cognitive impairment [3]; in addition, if the observation is extended beyond 20 years, many "benign" forms develop disabilities [4]. This data has been supported by another prospective study demonstrating that 29% of patients with benign MS significantly worsen over a five-year follow-up period [5].

Other particular forms, such as transitional MS, are also described. This term is used to define two different conditions: a disease characterized by a progressive course that begins months or years after a clinical episode suggestive of MS, and the period characterized by a lack of disease activity between relapsing-remitting and secondary-progressive phase.

About 80% of patients present a mono-symptomatic acute onset, called Clinically Isolated Syndrome (CIS), generally characterized by the involvement of a single functional system. Even if it is not possible to identify symptoms or signs revealing the onset, nevertheless some of them more frequently characterize the first episode. It must also be considered that, although less frequently than in the past, a hypothetical first episode of MS is still reconstructed based on a description by the patient and/or his family members of a more or less remote pathological event initially interpreted as an irrelevant event or not related to a possible onset of MS.

In most cases the onset symptoms may be weakness of one or more limbs (40%), optic neuritis (22%), sensory impairments such as paresthesias and dysesthesias (18%), cerebellar symptoms (15%); more rarely onset symptoms are characterized by diplopia, vertigo and bladder dysfunctions (10%), paroxysmal or psychiatric symptoms (5%). Some patients (20%) show a polysymptomatic onset [6].

Onset symptoms vary considerably according to age; when the onset occurs at an advanced age sensitivity and motor impairments are likely to be more frequent, while oculomotor disturbances and visual impairment such as optic neuritis usually occur in younger patients. The frequency of cerebellar disorders is not affected by age at onset of MS.

Although the neurological symptoms and signs mentioned above, both as isolated and variously combined syndromes, are fairly frequent at disease onset, there are many other deficits involving the CNS which may still be indicative of MS, even though they show more low frequencies at onset. When a specific neurological symptom or sign does not clearly suggest the diagnosis of MS, instrumental tests and in particular MRI examination can often raise a suspicion of the disease.

Other onset symptoms and signs may rarely include: isolated involvement of cranial nerves other than the optic nerve and oculomotor nerves (i.e. trigeminal neuralgia or facial palsy); paroxysmal symptoms and epileptic events, cognitive function impairment, psychiatric disorders, sphincter and sexual dysfunction. All these features show a higher frequency, sometimes considerable, in the course of the disease compared to the onset.

New episodes randomly occur with a relapse rate of about 1.5 per year. In the early stages of the disease there is usually a full recovery from symptoms after every attack; later, recovery tends to be partial with accumulation of disability. Many studies have been performed on clinical or demographic variables that can influence the course of disease and survival in MS [7–10]. Age of onset, sex and symptoms at onset are the most widely studied prognostic factors. These variables have been shown to be interdependent. Initial symptoms of optic neuritis or sensory dysfunction have been shown to be favorable prognostic factors [11–13], while pyramidal and cerebellar symptoms, and the presence of several different symptoms, appear to be associated with a worse prognosis [14].

Due to the random wide spatial distribution of lesions, MS can produce symptoms and signs of CNS involvement in all functional systems, although some disorders are common and others are rare.

3.1 Pyramidal System Dysfunction

Almost 100% of patients show clinical or subclinical involvement of the pyramidal system. Hyposthenia of one or more limbs, characterized by variable degrees of weakness from mild paresis to plegia, can be present. Hyposthenia is often associated with spasticity, which can occur both in flexion and extension. Spasticity is a major cause of disability: it interferes with the physiological movement of the limbs and can lead to tendon alterations and ankylosis of the affected joints; in severe forms it may involve the axial muscles, interfering with normal respiratory function and forcing the patient to assume awkward postures, creating pathological conditions. Neurological examination shows clear signs of central spastic paresis with increased deep tendon reflexes, pathological reflexes such as Hoffmann's and Babinski's signs, ankle clonus and, more rarely, knee clonus.

3.2 Sensory System Dysfunction

Sensory impairment is often the presenting symptom and regularly occurs during the course of the disease. It is caused by lesions localized in posterior columns, spinothalamic tracts or the dorsal root entry zone. The patchy distribution of the reduction or loss of tactile and/or heat and pain sensitivity is a typical feature, but these symptoms, like all other sensory deficits, do not show any distinctive features compared to the same dysfunctions occurring in other diseases, apart from their irregular distribution. They are reported by patients as feelings of numbness, tingling, wrapping or swelling. On physical examination, especially in the early stages, there are no marked deficits, but the patient may experience tactile, heat or pain stimulation as remote or different. In more advanced stages paresthesia and dysesthesia in the fingers and toes and distal hypoesthesia in the lower limbs often remain, and an impairment of deep sensitivity, particularly pallesthesia, in the lower limbs is very often present. Many patients describe a sensation of "electric shock" or tingling on neck flexion localized in the trunk near paravertebral muscles, extending down the lower limbs. This symptom, called Lhermitte's sign, has been considered almost pathognomonic of MS; however, it can also occur as a result of cervical spinal cord injury or meningeal irritation due to other etiology.

3.3 Cerebellar System Dysfunction

At onset cerebellar symptoms are not frequent, but they may become common later. They may appear subtly, perceived as dizziness accompanied by subjective vertigo, instability and disequilibrium. Only in the most severe cases may static and dynamic ataxia with spastic ataxic gait, intention tremor, dysdiadochokinesia, hypotonia, scanning speech and nystagmus appear. Nystagmus is due to cerebellar tract or internuclear brainstem pathway lesions. These symptoms are very disabling, with poor tendency to regression and resistant to available symptomatic treatments, so that their occurrence from the onset of MS consistently suggests a worse prognosis.

3.4 Visual System Dysfunction

Optic neuritis is the presenting symptom in 22% of cases and over time about 50% of patients developing an optic neuritis convert to clinically definite MS. In young MS patients its incidence is currently estimated at about 20–25% of cases, and appears to decrease progressively with age. Clinical signs and symptoms suggestive of optic neuritis include acute or subacute unilateral (rarely bilateral) deficits of visual function, often associated with retrobulbar pain (exacerbated by eye movements). The decrease of visual acuity (from a few degrees to complete blindness) is not correctable with lenses, and is generally perceived by the patient as blurred vision. The decrease of visual acuity becomes even more severe with increased temperature or fever or prolonged exercise. This phenomenon, evidence of a transient worsening of nerve conduction, is clinically defined as "Uhtoff's phenomenon". However, there are asymptomatic forms, diagnosis of which can be made based on an increase in Visual Evoked Potential (VEP) latency. Generally, the rapid worsening of visual function becomes less severe within 10–14 days; then there is a gradual improvement during the next 4–6 weeks, although sometimes recovery may be partial and decreased color sensitivity remains. During the acute phase the optic disc is almost always normal, but takes on a pale appearance in a few weeks after onset. This pallor is typically temporal and remains stable even on full recovery. The characteristic visual field defect is represented by a central scotoma, although, based on an American multicenter study [10], bilateral deficits, paracentral and arcuate scotomas may also be found.

3.5 Pain

Various types of pain may occur due to MS; their overall incidence is quite high, and this symptom may be underestimated. In addition to paroxysmal pain (e.g. neuralgia), chronic neuropathic and musculoskeletal pain are described: the first most often affects the lower limbs and occurs with variable features, while the latter is due to several different factors.

3.6 Brainstem Dysfunction

This definition may also include motor and cerebellar dysfunctions; however, these symptoms are considered separately, because they are due to lesions of specific neurological systems. Brainstem dysfunction refers to bulbar palsy or disorders affecting cranial nerves and the vestibular system.

When the medulla oblongata is affected, symptoms such as dysphagia and dysarthria or abnormal sensitivity in the mouth and throat manifest, whereas damage involving the cortical-bulbar fibers causes the so-called pseudo-bulbar syndrome. In addition to dysarthria and dysphagia due to supranuclear damage, this syndrome is characterized by emotional incontinence and uncontrolled outbursts of laughing and crying.

A lesion of the sixth, third or, more rarely, fourth cranial nerve may be responsible for the onset of diplopia, often with a favorable clinical course and regression. Inter-nuclear ophthalmoplegia is a more common symptom caused by longitudinal fasciculus lesions, manifesting as an adduction deficit on the side of the lesion with dissociated nystagmus of the abducting eye.

Facial palsy, sensory deficits caused by damage to the trigeminal nerve, and nystagmus are frequently described.

3.7   Autonomic Dysfunction

About 90% of MS patients complain of bladder, bowel or sexual dysfunction during the course of the disease.

3.7.1   Bladder Dysfunction

This can present as three different patterns: hyperactive or uninhibited bladder; hypoactive or flaccid bladder; detrusor-sphincter dyssynergia.

One important consequence of bladder dysfunction is the high risk of urinary tract infections with the possibility of generalized infections (sepsis of urinary origin), and kidney involvement, which may lead to impaired renal function.

Lower limb spasticity affects bladder contraction; in fact, there are significant correlations between the degree of motor dysfunction caused by pyramidal system damage and the severity of bladder dysfunction. Bladder and urinary tract infections negatively affect pyramidal hypertonia in the lower limbs.

3.7.2   Bowel Dysfunction

Constipation is more common, but fecal urgency and incontinence can also occur; probably a number of factors related to dysfunctions of the internal and external anal sphincters as well as to food and liquid intake, and to psychological and management problems, contribute to these disorders.

3.7.3   Sexual Dysfunction

Sexual dysfunction seems to be more frequent in males than in females; in men erectile dysfunction is the most common, with variable severity, but ejaculation and orgasmic dysfunction can also occur. In women anorgasmy, vaginal dryness and decreased libido are reported. The etiology of sexual dysfunction may be primarily organic (perineal sensory impairment, alterations in autonomic innervation), related to other physical deficits (movement and muscle tone disorders involving lower limbs, bladder and bowel dysfunction, fatigue) but it may also results from psychological problems (mood disorders, embarrassment, loss of interest and self-esteem).

3.8   Cognitive Dysfunction and Psychiatric Disorders

These are important clinical features of MS patients due to their frequency and impact on functional capacity. In keeping with the specific topic of this text, these aspects will be discussed in detail in the relevant sections.

3.9   Disorders Difficult to Diagnose and Categorize

3.9.1   Fatigue

Fatigue related to MS is defined as a feeling of tiredness not proportional to the effort exerted, or a feeling of weakness and inability to produce sufficient muscle strength, or the inability to sustain physical or mental performance. Fatigue was defined by the Multiple Sclerosis Council for Clinical Practical Guidelines [15] as "a subjective lack of physical and/or mental health that is perceived by the individual or caregiver to interfere with usual and desired activities". Regardless of how it is defined, fatigue (or fatigability) is a very common disorder in MS patients; in several cases it significantly affects functional efficiency, and many patients regard this disorder as the most problematic of the symptoms they note [16–18].

Fatigue is often already present at onset and persists during the disease course, characterized by a pattern of increasing and decreasing intensity which is sometimes unexplained and sometimes connected to a relapse or other intercurrent events causing an increase in body temperature (e.g., physical exercise, especially if excessive, or increase of ambient temperatures). Fatigue intensity varies from mild to severe and does not seem to be significantly correlated with the degree of disability. It usually shows temporal variations related to time of day, being milder in the morning (at first awakening) or after rest periods, and more severe in the evening.

3.9.2   Headache

Headache is frequent in MS patients, but this phenomenon has been interpreted as a coincidence due to the high frequency of the symptom. Some types of headache, for example the tension type, are more frequent, and seem to be related to cervical or scalp muscle spasms. In other cases, the headache is due to large tumor-like lesions or lesions located in areas involved in CSF dynamics resulting in a blockage of CFS circulation and an increase in intracranial pressure. In other cases the symptom may be related to retrobulbar optic neuritis or oculomotor disorders.

3.9.3   Paroxysmal Symptoms

Symptoms that occur suddenly, for a short period of time (from a few seconds to 2–3 min) and then spontaneously disappear are included in this category. They may even occur dozens or hundreds of times a day, and this situation usually continues for a period of time ranging from several weeks to several months. They affect a significant percentage of patients, but an unequivocal correlation with specific areas of damage does not seem to exist, while there seems to be a correlation with elevated body temperature. Clinical and electroencephalographic data rule out an epileptic origin of these symptoms; nevertheless, we should not be surprised at the efficacy of antiepileptic drugs in controlling these disorders. The causal mechanism of paroxysmal disorders is very likely due to changes induced by the demyelinating process affecting the electrochemical properties of axonal membranes. Paroxysmal disorders may include episodes of diplopia, ataxia, dysarthria, tingling, spasms and involuntary movements, transient lower limb weakness, pain, dysesthesia and paresthesia.

3.10 Other Disorders

3.10.1 Seizures

Their incidence is rather low; the cause is often not related to MS, but to other intercurrent disease. Among MS lesions, those located at the gray-white matter junction (juxtacortical lesions) are more frequently linked to the genesis of seizures.

3.10.2 Sleep Disorders

These disorders (insomnia, nocturnal movement disorders, sleep-disordered breathing, narcolepsy and REM sleep disorders) seem to have a rather high incidence; sleep disorders are caused by several different factors, and for most of them it is not possible to distinguish if they are a direct (particular location of lesions) or indirect consequence of the disease.

References

1. Lublin FD, Reingold SC (1996) Defining the clinical course of multiple sclerosis: results of an International survey. Neurology 46:907–911
2. Bashir K, Whitaker JN (2002) Handbook of multiple sclerosis. Lippincot Williams & Wilkins, Philadelphia
3. Amato MP, Zipoli V, Goretti B et al (2006) Benign multiple sclerosis: cognitive, psychological and social aspects in a clinical cohort. J Neurol 253:1054–1059
4. Sayao AL, Devonshire V, Tremlett H (2007) Longitudinal follow-up of "benign" multiple sclerosis at 20 years. Neurology 68:496–500
5. Portaccio E, Stromillo ML, Goretti B et al (2009) Neuropsychological and MRI measures predict short-term evolution in benign multiple sclerosis. Neurology 73:498–503
6. Coles AJ, Compston A (2004) Multiple sclerosis. Medicine 32:87–92
7. Hyllested K (1961) Lethality, duration, and mortality of disseminated sclerosis in Denmark. Acta Psychiatr Scand 36:553–564
8. Leibowitz U, Alter M (1970) Clinical factors associated with increased disability in multiple sclerosis. Acta Neurol Scand 46:53–70
9. Poser S, Raun NE, Poser W (1982) Age at onset, initial symptomatology and the course of multiple sclerosis. Acta Neurol Scand 66:355–362
10. Visscher BR, Liu KS, Clark VA, Detels R, Malmgren RM, Dudley JP (1984) Onset symptoms as predictors of mortality and disability in multiple sclerosis. Acta Neurol Scand 70:321–328

11. Weinshenker BG, Bass B, Rice GP et al (1989) The natural history of multiple sclerosis: a geographically based study. 2. Predictive value of the early clinical course. Brain 112:1419–1428

12. Rodriguez M, Siva A, Ward J, Stolp-Smith K, O'Brien P, Kurland L (1994) Impairment, disability, and handicap in multiple sclerosis: a population-based study in Olmsted County, Minnesota. Neurology 44:28–33

13. Pittock SJ, Mayr WT, McClelland RL et al (2004) Change in MS-related disability in a population-based cohort: a 10-year follow-up study. Neurology 62:51–59

14. Optic Neuritis Study Group (1991) The clinical profile of optic neuritis: experience of the Optic Neuritis Treatment Trial. Arch Ophtalmol 109:1673–1678

15. Multiple Sclerosis Council for Clinical Practice Guidelines (1998) Fatigue and multiple sclerosis: Evidence-based management strategies for fatigue in multiple sclerosis. Paralyzed Veterans of America, Washington, DC

16. Krupp LB, Alvarez LA, LaRocca NG, Scheinberg LC (1988) Fatigue in multiple sclerosis. Arch Neurol 45:435–437

17. Bergamaschi R, Romani A, Versino M, Poli R, Cosi V (1997) Clinical aspects of fatigue in multiple sclerosis. Funct Neurol 12:247–251

18. Fisk JD, Pontefract A, Ritvo PG, Archibald CJ, Murray TJ (1994) The impact of fatigue on patients with multiple sclerosis. Can J Neurol Sci 21:9–14

Etiopathogenesis

4

Silvia Romano, Carlo Caltagirone, and Ugo Nocentini

The etiology of MS is unknown, as are largely unknown the mechanisms responsible for the pathogenesis of demyelination and the clinical course. However, experimental models of the disease, such as experimental autoimmune (or allergic) encephalomyelitis (EAE), a virus-induced demyelination, and a range of clinical and laboratory data, suggest that MS is a multi-factorial disease in which one or more environmental factors contribute to triggering an immune response directed against myelin antigens in genetically susceptible subjects.

From a pathogenic point of view it is important to consider:

(a) Genetic factors: there are many studies that have attempted to analyze the genetic aspects of this disease. MS is characterized by a marked familial tendency; the relatives of patients appear to have an increased risk of developing the disease about 20–50 times higher than those who do not have an affected family member. Studies on monozygotic twins have also confirmed concordance values from 25 % to 31 % [1], which are about 6 times the concordance rate in dizygotic twins (5 %). As early as the 70s MS was known to be associated with MHC alleles [2, 3]; more recent studies have confirmed this association, identifying a susceptibility linked to the DR15 and DQ6 haplotypes and the corresponding DRB1 * 1501, DRB5 * 0101, DQA1 * 0102, * 0602 and DQB2 genotypes [4]. This association is present in all populations and is stronger in Northern Europe. However, some exceptions are described, such as the populations of Sardinia and some Mediterranean areas where an

S. Romano (✉)
Centre for Experimental Neurological Therapies (CENTERS), Neurology Unit, S. Andrea Hospital, Rome, Italy
e-mail: silvia.romano@uniroma1.it

C. Caltagirone • U. Nocentini
Dipartimento di Neuroscienze, Università degli Studi di Roma "Tor Vergata", Rome, Italy
e-mail: c.caltagirone@hsantalucia.it; u.nocentini@hsantalucia.it

U. Nocentini et al. (eds.), *Neuropsychiatric Dysfunction in Multiple Sclerosis*,
DOI 10.1007/978-88-470-2676-6_4, © Springer-Verlag Italia 2012

association with the DR4 haplotype (DRB1 * 0405-DQA1 * 0301-DQB1 * 0302) has been identified [5].

These results have prompted the development of new approaches to identify other genetic risk factors: association and linkage studies of candidate genes based on the knowledge of disease pathogenesis; genome screening studies analyzing specific chromosomal regions and using markers to create "linkage disequilibrium" studies; studies on genetically isolated populations, such as the population of Sardinia, which are ideal for a better understanding of the role of genetic factors.

These new methodological approaches have led to the discovery of a possible association with the gene encoding interleukin-2 receptor and interleukin-7 receptor alpha chain [6–9]; considerable evidence for several new MS susceptibility loci (KIF21B and TMEM39A) has recently been presented by the International Multiple Sclerosis Genetics Consortium (IMSGC) [10].

(b) Viral factors: epidemiological studies have suggested the existence of one or more environmental factors that may play a role during childhood and adolescence and for which subjects who go on to develop MS seem to have a particular susceptibility. Many infectious agents, in particular viral agents, have been involved in the disease, but none has been clearly linked to it. Studies using an animal model of MS, the EAE, have revealed that an autoimmune response directed to self antigens is behind the development of the disease. The animal model has highlighted the fact that the basic requirement to begin the autoimmune process leading to CNS demyelination is the activation of myelin antigen-specific T cells in the peripheral blood [11, 12]. This process could be triggered by molecular mimicry between viral proteins and self CNS antigens, as well as by phenomena of bystander activation (the nonspecific activation of autoreactive cells due to inflammatory events caused by infection). The most studied viruses include some herpes viruses causing a chronic infection in humans, the neurotropic human herpes virus 6 (HHV6) and the lymphotropic Epstein-Barr virus (EBV). While serological and molecular data correlating the virus HHV6 with the development of the disease appear to be in disagreement [13], the results related to EBV are more reliable. In fact, it has been demonstrated that the level of virus seropositivity is significantly higher in patients than in healthy subjects [14]; moreover, a recent study has shown the presence of EBV-infected B cells in lymphoid follicle-like structures in the cerebral meninges of some MS patients, suggesting a correlation among the reactivation of the virus in the follicles, acute inflammation and relapses [15].

(c) Immunological factors: clinical, anatomo-pathologic and experimental data support the hypothesis that MS is an autoimmune inflammatory disease. In fact, an involvement of both T and B lymphocytes has been demonstrated. T lymphocytes play a key role in adaptive immunity (i.e., in response to specific antigens already known) by acting directly against the cells displaying foreign antigens, such as virus or cancer cells (cytotoxic T lymphocytes), by producing cytokines required for the proliferation of other immune system cells (T helper1 and T helper2 lymphocytes) or by regulating the immune responses (suppressor and regulatory T lymphocytes). T lymphocytes are commonly classified as CD4 and CD8 T cells according to the presence of surface antigens; helper and regulatory T cells belong to the first group, while the second group consists of cytotoxic and suppression T lymphocytes. On the other hand, the functions of B cells are to produce antibodies and present antigens to T helper cells.

In rheumatoid arthritis as well as MS, the immune response tends to be primarily mediated by T helper 1 rather than T helper 2 cells, with the production of cytokines such as interferon gamma and interleukin-2 [15]. This situation is confirmed by clinical worsening following the administration of interferon-γ, suggesting its probable pro-inflammatory action [16].

The hypothesis of autoimmune pathogenesis is also supported by the existence of an EAE model induced by immunizing susceptible animals with myelin-derived antigens and mediated by MBP specific CD4+ T cells [11, 12]; transferring these cells from affected animals into healthy control animals has been shown to cause the development of the disease. EAE has been extremely useful for studying some pathogenic mechanisms underlying MS, due to the several anatomo-pathologic and clinical similarities between these diseases, such as lymphomonocytes infiltration, gliosis and the demyelination process.

In recent years several studies have focused on the action of regulatory cells, demonstrating that in MS patients there might be a hypofunction of a subset of these cells (CD4+ CD25) [17, 18].

Alterations of antibody-mediated immune response are also described: about 90 % of MS patients show oligoclonal bands in CSF and for many years their presence has been essential to meet the diagnostic criteria for MS [19]. Many studies have identified serum antibodies directed against myelin protein: Berger and colleagues [20] detected the presence of anti-MOG (Myelin Oligodendrocyte Glycoprotein) antibodies associated or not associated to anti-MBP (Myelin Basic Protein) antibodies in patients with a first clinical episode and the presence of multifocal lesions on MRI suggestive of MS. These antibodies seem to affect the course of the disease because their presence represents a risk factor for the conversion to definite MS.

The exact cause of MS remains unknown. As has been demonstrated, it is generally agreed that abnormal immune mechanisms play an important role in causing the typical lesions of the disease; many data about the characteristics of these processes result from EAE studies. Some researchers periodically point out that caution is always required when automatically transferring information derived from EAE to human disease [see 21, for a recent discussion of the issue].

Many researchers cast doubt on the classification of MS as an autoimmune disease [22]. Recently, Zamboni and colleagues have given new support to the vascular theory of MS, originally proposed over a century ago, suggesting that chronic cerebrospinal venous insufficiency (CCSVI) plays a role in causing MS. CCSVI is likely to cause obstructed venous flow leading to tissue iron accumulation, which triggers and maintains inflammation in CNS [23, 24]. Zamboni has also described symptomatic relief after what some call the 'liberation procedure' (a procedure involving angioplasty or stenting) of certain veins in an attempt to improve blood flow [25, 26]. The "liberation procedure" has been criticized for possibly resulting in serious complications and deaths while its benefits have not been proven. Accordingly, within the medical community, both CCSVI and this procedure have been met with skepticism. Zamboni and colleagues initially reported 100 % sensitivity, specificity and positive and negative predictive values for detecting CCSVI in MS patients [27] but no study, since this report, has yielded similar results, while a great variability of CCSVI has been found in both MS patients and in control subjects [28–32]. This wide variability is explained by methodological aspects, problems in assessing CCSVI, and differences among clinical series. Moreover, some abnormalities were reported in other neurological conditions such as transient global amnesia and neurodegenerative diseases [33–35]. Additional studies using an appropriate epidemiological methodology to define the possible relationship between CCSVI and MS are underway. However, this hypothesis reminds us of the need to explore every possible aspect in the search for a correct pathogenic interpretation, without closure and with the awareness that a paradigm shift may sometimes be required [36].

References

1. Dyment DA, Ebers GC, Sadovnick AD (2004) Genetics of multiple sclerosis. Lancet Neurol 3:104–110
2. Compston DA, Batchelor JR, McDonald WI (1976) B-lymphocyte alloantigens associated with multiple sclerosis. Lancet 308:1261–1265
3. Terasaki PI, Park MS, Opelz G, Ting A (1976) Multiple sclerosis and high incidence of a B lymphocyte antigen. Science 193:1245–1247
4. Olerup O, Hillert J (1991) HLA class II-associated genetic susceptibility in multiple sclerosis: a critical evaluation. Tissue Antigens 38:1–15
5. Marrosu MG, Muntoni F, Murru MR et al (1992) HLA-DQB1 genotype in Sardinian multiple sclerosis: evidence for a key role of DQB1 *0201 and *0302 alleles. Neurology 42:883–886
6. Gregory SG, Schmidt S, Seth P et al (2007) Interleukin 7 receptor alpha chain (IL7R) shows allelic and functional association with multiple sclerosis. Nat Genet 39:1083–1091
7. International Multiple Sclerosis Genetics Consortium, Hafler DA, Compston A, Sawcer S et al (2007) Risk alleles for multiple sclerosis identified by a genomewide study. N Engl J Med 357:851–862
8. Lundmark F, Duvefelt K, Iacobaeus E et al (2007) Variation in interleukin 7 receptor alpha chain (IL7R) influences risk of multiple sclerosis. Nat Genet 39:1108–1113
9. International Multiple Sclerosis Genetics Consortium (IMSGC) (2008) Refining genetic associations in multiple sclerosis. Lancet Neurol 7:567–569
10. International Multiple Sclerosis Genetics Consortium (IMSGC) (2010) Comprehensive follow-up of the first genome-wide association study of multiple sclerosis identifies KIF21B and TMEM39A as susceptibility loci. Hum Mol Genet 19:953–962
11. Goverman J, Woods A, Larson L, Weiner LP, Hood L, Zaller DM (1993) Transgenic mice that express a myelin basic protein-specific T cell receptor develop spontaneous autoimmunity. Cell 72:551–560
12. Moore FG, Wolfson C (2002) Human herpes virus 6 and multiple sclerosis. Acta Neurol Scand 106:63–83
13. Wandinger K, Jabs W, Siekhaus A et al (2000) Association between clinical disease activity and Epstein-Barr virus reactivation in MS. Neurology 55:178–184

14. Serafini B, Rosicarelli B, Franciotta D et al (2007) Dysregulated Epstein-Barr virus infection in the multiple sclerosis brain. J Exp Med 204:2899–2912
15. Lafaille JJ (1998) The role of helper T cell subsets in autoimmune diseases. Cytokine Growth Factor Rev 9:139–151
16. Panitch HS (1992) Interferons in multiple sclerosis. A review of the evidence. Drugs 44:946–962
17. Feger U, Luther C, Poeschel S, Melms A, Tolosa E, Wiendl H (2007) Increased frequency of CD4+ CD25+ regulatory T cells in the cerebrospinal fluid but not in the blood of multiple sclerosis patients. Clin Exp Immunol 147:412–418
18. Venken K, Hellings N, Broekmans T, Hensen K, Rummens JL, Stinissen P (2008) Natural naive CD4 + CD25 + CD127 low regulatory T cell (Treg) development and function are disturbed in multiple sclerosis patients: recovery of memory Treg homeostasis during disease progression. J Immunol 180:6411–6420
19. Freedman MS, Thompson EJ, Deisenhammer F et al (2005) Recommended standard of cerebrospinal fluid analysis in the diagnosis of multiple sclerosis: a consensus statement. Arch Neurol 62:865–870
20. Berger T, Rubner P, Schautzer F et al (2003) Antimyelin antibodies as a predictor of clinically definite multiple sclerosis after a first demyelinating event. N Engl J Med 349:139–145
21. Codarri L, Fontana A, Becher B (2010) Cytokine networks in multiple sclerosis: lost in translation. Curr Opin Neurol 23:205–211
22. Chauduri A, Behan PO (2004) Multiple Sclerosis is not an autoimmune disease. Arch Neurol 61:1610–1612
23. Zamboni P (2006) The big idea: iron-dependent inflammation in venous disease and proposed parallels in multiple sclerosis. J R Soc Med 99:589–593
24. Zamboni P, Galeotti R, Menegatti E et al (2009) A prospective open-label study of endovascular treatment of chronic cerebrospinal venous insufficiency. J Vasc Surg 50:1348–1358
25. Zamboni P, Galeotti R, Weinstock-Guttman B, Kennedy C, Salvi F, Zivadinov R (2012) Venous angioplasty in patients with multiple sclerosis: results of a pilot study. Eur J Vasc Endovasc Surg 43:116–122
26. Zamboni P, Galeotti R, Menegatti E et al (2009) Chronic cerebrospinal venous insufficiency in patients with multiple sclerosis. J Neurol Neurosur Ps 80:392–399
27. Singh AV, Zamboni P (2009) Anomalous venous blood flow and iron deposition in multiple sclerosis. J Cereb Blood Flow Metab 29:1867–1878
28. Doepp F, Paul F, Valdueza JM et al (2010) No cerebrocervical venous congestion in patients with multiple sclerosis. Ann Neurol 68:173–183
29. Sundsrom P, Wahlin A, Ambarki K et al (2010) Venous and cerebrospinal fluid flow in multiple sclerosis: a case–control study. Ann Neurol 68:255–259
30. Baracchini C, Perini P, Calabrese M, Causin F, Rinaldi F, Gallo P (2011) No evidence of chronic cerebrospinal venous insufficiency at multiple sclerosis onset. Ann Neurol 69:90–99
31. Baracchini C, Perini P, Causin F, Calabrese M, Rinaldi F, Gallo P (2011) Progressive multiple sclerosis is not associated with chronic cerebrospinal venous insufficiency. Neurology 77:844–850
32. Centonze D, Floris R, Stefanini M et al (2011) Proposed chronic cerebrospinal venous insufficiency criteria do not predict multiple sclerosis risk or severity. Ann Neurol 70:51–58
33. Sander D, Winbeck K, Etgen T et al (2000) Disturbance of venous flow patterns in patients with transient global amnesia. Lancet 356:1982
34. Wattjes MP, Oosten BW, de Graaf W et al (2011) No association of abnormal cranial venous drainage with multiple sclerosis: a magnetic resonance venography and flow-quantification study. J Neurol Neurosurg Psychiatry 82:429–435
35. Zivadinov R, Marr K, Cutter G et al (2011) Prevalence, sensitivity, and specificity of chronic cerebrospinal venous insufficiency in MS. Neurology 77:138–144
36. Barnett MH, Sutton I (2006) The pathology of multiple sclerosis: a paradigm shift. Curr Opin Neurol 19:242–247

Neuropathology

5

Silvia Romano, Carlo Caltagirone, and Ugo Nocentini

In dealing with the neuropathological aspects of MS, we will refer to the classic form of the disease. The typical histopathological lesion of MS is the demyelinating plaque from which the alternative name "sclérose en plaques" derives. Demyelinating plaques vary in size and shape.

It is well established that plaques are present not only in the white matter of the brain, cerebellum, brainstem and spinal cord, but also in the gray matter. The areas of white matter most commonly involved in inflammation and demyelination processes, where the plaques are easily identifiable, are the periventricular regions, optic nerve, corpus callosum and spinal tracts.

Several classifications of the plaques have been proposed, based on the activity level of pathological processes, disease stage and supposed pathogenetic processes.

One possible classification divides the plaques into:

- Active plaques: characterized by perivascular lymphocytes and macrophage infiltrates. Several macrophages with sizeable lipid inclusions are distributed within the plaque; these lipid inclusions are the result of macrophage activation against myelin debris.
- Inactive plaques: characterized by low cellularity and severe gliosis associated with axonal loss.
- Chronic active plaques: these lesions also show low cellularity and intense gliosis but associated with lipid-laden macrophages in the periphery of the plaque.

S. Romano (✉)
Centre for Experimental Neurological Therapies (CENTERS), Neurology Unit, S. Andrea Hospital, Rome, Italy
e-mail: silvia.romano@uniroma1.it

C. Caltagirone • U. Nocentini
Dipartimento di Neuroscienze, Università degli Studi di Roma "Tor Vergata", Rome, Italy
e-mail: c.caltagirone@hsantalucia.it; u.nocentini@hsantalucia.it

U. Nocentini et al. (eds.), *Neuropsychiatric Dysfunction in Multiple Sclerosis*,
DOI 10.1007/978-88-470-2676-6_5, © Springer-Verlag Italia 2012

– Shadow plaques: defined as sharply circumscribed areas characterized by myelin loss and remyelination in which axons are surrounded by thin myelin sheaths.

The CNS sections show the concurrent presence of different types of plaques that may be distinguished from each other by degree of activity, shape or size. Cerebral, cerebellar and brainstem involvement is manifestly asymmetrical. The plaques can also be found in the cerebral cortex, especially in the subpial region, deep nuclei and spinal cord gray matter.

Although the first neuropathologic studies on MS already described the involvement of gray matter [1–6], it is only in recent years, due to new MRI techniques, that it has been possible to determine the importance of gray matter lesions. The introduction of *double inversion recovery* MRI sequences, which selectively suppress the CSF and white matter signal, has made it possible to show that cortical lesions are consistently present, not only in patients with a progressive disease but also in patients with CIS and RR MS [7, 8]. According to the morphological classification, cortical lesions are divided into: (1) cortical-juxtacortical lesions extending into the deeper cortical laminae and the subcortical white matter (type I lesions), (2) lesions extending through all cortical layers without involving the white matter and often characterized by the presence of a central vessel (type II lesions), and subpial lesions that extend from the pia mater to the cerebral cortex involving the most superficial cortical layers (type III/IV lesions) [9, 10]. Cortical lesions have been described more frequently in areas with a slower CSF circulation, such as the deeper grooves, insula and cingulate cortex [8–10]; these data support the hypothesis that a soluble factor produced by lymphocytic infiltrates in the meninges [11] can spread to the cerebral cortex, triggering the demyelination process, both directly, and indirectly by microglial activation. From an immunological point of view, compared with typical white matter lesions, cortical lesions show a less extensive inflammatory process and are characterized by a lower concentration of inflammatory cells (CD3+ and CD68+) [7], almost total absence of complement factor deposition [12] and integrity of the blood-brain barrier [13].

Histological examination makes it possible to identify other distinguishing characteristics of MS lesions compared with other myelin lesions: in MS plaques, spared axons are always detectable in regions where myelin damage has occurred; likewise, neuron bodies can be identified within the plaques both in cortical and deep gray matter. Perivenular localization of the plaques is also common in ADEM, but the features of perivenular infiltrates are different from those observed in MS.

In active lesions, the inflammation is supported by infiltrates of CD4+ and CD8+ T lymphocytes.

In inactive lesions, even B cells can rarely be detected. Phagocytosis of myelin debris is carried out by both blood-derived macrophage and activated resident microglia. In the oldest lesions, the cellularity, as mentioned above, is low, and lymphocytes, both T and B, can mainly be detected near the vessels. Microglial cells are abundant but not activated. Mononuclear infiltrates can be found even in the meninges. The presence of follicle-like structures in the meninges suggests that B lymphocyte maturation can occur in the cerebral compartment.

In the acute phase demyelination processes can occur with or without oligodendroglial cell loss. These data suggest that the mechanisms of demyelination may be manifold, i.e., demyelination may be mediated by antibodies and complement or T cells and macrophages; in other cases primary damage to oligodendrocytes may occur. The pattern of demyelination seems to change among patients, with some patients showing constant patterns over time.

Due to new radiological and histopathological data, axonal damage is becoming more and more important as a determining factor in the functional impairment of MS patients. Furthermore, axonal injury is characterized by the accumulation of amyloid precursor protein, but the heterogeneity also seems to involve the pathogenesis of this damage.

Another issue complicating the histological pattern of MS is the presence of remyelination processes, which seem to vary greatly among patients; the extent of remyelination also appears to differ between cortical and brain white matter lesions.

References

1. Dawson JW (1962) The histology of multiple sclerosis. Trans R Soc Edinburgh 50:517–740
2. Brownell B, Hughes JT (1962) The distribution of plaques in the cerebrum in multiple sclerosis. J Neurol Neurosurg Psychiatry 25:315–320
3. Lumsden CE (1970) The neuropathology of multiple sclerosis. In: Vinken PJ, Bruin GW (eds) Handbook of Clinical Neurology. Elsevier Science Publishers, Amsterdam, pp 217–309
4. Calabrese M, De Stefano N, Atzori M et al (2007) Detection of cortical inflammatory lesions by double inversion recovery magnetic resonance imaging in patients with multiple sclerosis. Arch Neurol 64:1416–1422
5. Calabrese M, Rocca MA, Atzori M et al (2009) Cortical lesions in primary progressive multiple sclerosis: a 2-year longitudinal MR study. Neurology 72:1330–1336
6. Kidd D, Barkhof F, McConnell R et al (1999) Cortical lesions in multiple sclerosis. Brain 122:17–26
7. Peterson JW, Bo L, Mork S et al (2001) Transected neurites, apoptotic neurons, and reduced inflammation in cortical multiple sclerosis lesions. Ann Neurol 50:389–400
8. Bo L, Vedeler CA, Nyland HI, Trapp BD, Mork S (2003) Subpial demyelination in the cerebral cortex of multiple sclerosis patients. J Neuropathol Exp Neurol 62:723–732
9. Kutzelnigg A, Lassmann H (2006) Cortical demyelination in multiple sclerosis: a substrate for cognitive deficits? J Neurol Sci 245:123–126
10. Kutzelnigg A, Lucchinetti CF, Stadelmann C et al (2005) Cortical demyelination and diffuse white matter injury in multiple sclerosis. Brain 128:2705–2712
11. Magliozzi R, Howell O, Vora A et al (2007) Meningeal B-cell follicles in secondary progressive multiple sclerosis associate with early onset of disease and severe cortical pathology. Brain 130:1089–1104
12. Brink BP, Veerhuis R, Breij EC, van der Valk P, Dijkstra CD, Bö L (2005) The pathology of multiple sclerosis is location-dependent: no significant complement activation is detected in purely cortical lesions. J Neuropathol Exp Neurol 64:147–155
13. Van Horssen J, Brink BP, de Vries HE, van der Valk P, Bø L (2007) The blood-brain barrier in cortical multiple sclerosis lesions. J Neuropathol Exp Neurol 66:321–328

The Diagnosis of Multiple Sclerosis

Simona Bonavita and Gioacchino Tedeschi

The diagnosis of Multiple Sclerosis (MS) is based on a careful evaluation of clinical history, neurological examination, and paraclinical tests aimed at identifying the temporal and spatial dissemination of demyelinating lesions of the Central Nervous System (CNS). On the other hand, it is mandatory to exclude any other disease which may better explain the clinical symptoms.

In 2001 diagnostic criteria [1] were elaborated to define the clinical and paraclinical (instrumental) evidence necessary for the diagnosis of MS. They were revised in 2005 [2] and more recently in 2010 [3]; instrumental examinations include Magnetic Resonance Imaging (MRI), cerebrospinal fluid (CSF) analysis and visual evoked potentials (VEP). MRI is necessary to demonstrate dissemination in time and space of demyelinating lesions. Before the final revision of diagnostic criteria, to hypothesize a diagnosis of MS, the brain MRI has to fulfill at least three of the four following criteria [4, 5]: (1) ≥ 1 gadolinium (Gd) enhancing lesion in T1 weighted images or ≥ 9 T2 hyperintense lesions; (2) ≥ 1 infratentorial lesion(s); (3) ≥ 1 subcortical lesion(s); and (4) ≥ 3 periventricular lesions.

One spinal lesion is equivalent to an infratentorial lesion and a Gd enhancing (Gd+) lesion in the spine is equivalent to a Gd + brain lesion. Moreover, any spine lesion, together with brain lesions, may contribute to reach the number of T2 hyperintense lesions necessary to satisfy the spatial dissemination criterion.

The recent revision of the McDonald criteria [3] recommends modifications of MRI criteria for dissemination in space (DIS). In the past version of the McDonald criteria, DIS demonstrated by MRI was based on the above-mentioned Barkhof/Tintoré criteria [4, 5]. Despite having good sensitivity and specificity, these criteria have been difficult to apply consistently by non-imaging specialists. The MAGNIMS multicenter collaborative research network suggests that DIS can be demonstrated with at least one T2 lesion in at least two of four locations considered characteristic

S. Bonavita (✉) • G. Tedeschi
Istituto di Scienze Neurologiche, Seconda Università di Napoli, Naples, Italy
e-mail: simona.bonavita@unina2.it

U. Nocentini et al. (eds.), *Neuropsychiatric Dysfunction in Multiple Sclerosis*,
DOI 10.1007/978-88-470-2676-6_6, © Springer-Verlag Italia 2012

for MS (juxtacortical, periventricular, infratentorial and spinal cord), with lesions within the symptomatic region excluded in patients with brainstem and spinal cord syndromes.

The 2005 revision of the McDonald criteria [2] based the dissemination in time (DIT) on the appearance of a new T2 lesion on a scan compared to a reference or baseline scan performed at least 30 days after the onset of the initial clinical event. In clinical practice, however, there is reason not to postpone a first MRI until after 30 days of clinical onset, as this would then require an extra MRI scan to confirm the diagnosis; the current revision of the McDonald criteria therefore allows a new T2 lesion to establish DIT, irrespective of the timing of the baseline MRI. More-over, the presence of both Gd + and non-enhancing asymptomatic lesions on the baseline MRI can substitute for a follow-up scan to confirm DIT, as long as it can be reliably determined that the Gd + lesion is not due to non-MS pathology.

CSF analysis may support a diagnosis of MS, in particular when the clinical presentation is atypical and/or the MRI criteria are not satisfied; moreover, it can exclude other diseases. CSF analysis is considered diagnostic when the IgG index is increased and two or more oligoclonal bands are present at isoelectric focusing. Lymphocytic pleocytosis is generally inferior to $50/mm^3$ and total protein concen-tration is inferior to 1 g/mm^3 [6].

VEP becomes necessary when either MRI or CSF analysis is insufficient to demonstrate spatial dissemination and to satisfy diagnostic criteria. They are considered positive when detecting an increase in P100 latency, with normal amplitude.

In 2005, it was established that the diagnosis of primary progressive MS (PPMS) required, in addition to 1 year of disease progression, two of the following three findings: positive brain MRI (nine T2 lesions or four or more T2 lesions with positive VEP); positive spinal cord MRI (at least two focal T2 lesions); positive CSF. In the 2010 revision the only novelty was the replacement of the previous brain imaging criteria with the new MRI criteria for DIS.

Therefore, MS diagnosis is based on the objective evidence of temporal (≥ 2 relapses) and spatial (≥ 2 lesions) dissemination. The simplest circumstance is when there are two clinical events (relapses) with objective evidence of two or more lesions of the CNS; this is the case of clinically definite MS. When there is evidence of less than two relapses and/or lesions, other paraclinical evidence is necessary to support the diagnosis. The most difficult situation is when there is a progressive neurological deficit suggestive of primary progressive MS; in this case more paraclinical evidence is necessary to make a diagnosis. When there is a single clinical event and a single objective lesion of the CNS, this is defined as a clinically isolated syndrome (CIS), which includes retrobulbar optic neuritis, transverse myelitis and brainstem syndrome. The risk of conversion to definite MS in the following 5 years is different in relation to the presence of MRI white matter lesions. In particular the risk is around 10 % in patients with negative MRI; it rises to around 50 % in the presence of 1–3 lesions and to 95 % when there are four or more lesions.

In summary, according to the 2010 McDonald Criteria for Diagnosis of MS, if the criteria are fulfilled and there is no better explanation for the clinical presentation, the diagnosis is "MS"; if the presentation is suspicious, but the criteria are not completely met, the diagnosis is "possible MS"; if during evaluation another diagnosis better explains the clinical presentation, then the diagnosis is "not MS."

An attack is defined as patient-reported or objectively observed events typical of an acute inflammatory demyelinating event in the CNS, current or historical, with duration of at least 24 h, in the absence of fever or infection. It should be documented by contemporaneous neurological examination, but some historical events, with symptoms and evolution characteristic for MS, for which no objective neurological findings are documented, can provide reasonable evidence of a prior demyelinating event. Reports of paroxysmal symptoms (historical or current) should, however, consist of multiple episodes occurring over not less than 24 h. Before a definite diagnosis of MS can be made, at least one attack must be corroborated by findings on neurological examination, VEP response in patients reporting prior visual disturbance, or MRI consistent with demyelination in the area of the CNS implicated in the historical report of neurological symptoms.

Clinical diagnosis based on objective clinical findings for two attacks is the most secure. Reasonable historical evidence for one past attack, in the absence of documented objective neurological findings, can include historical events with symptoms and evolution characteristics for a prior inflammatory demyelinating event; at least one attack, however, must be supported by objective findings, while no additional tests are required. However, any diagnosis of MS should be made with access to imaging based on the McDonald Criteria. If imaging or other tests (for instance, CSF) are undertaken and are negative, extreme caution is required before making a diagnosis of MS, and alternative diagnoses must be considered.

The 2010 revision of the McDonald Criteria can also be applied to paediatric patients, especially those with acute demyelination presenting as CIS, because most paediatric patients will have more than two lesions, and are very likely to have lesions in two of the four specified CNS locations. However, approximately 15–20 % of paediatric MS patients, most aged below 11 years, present with encephalopathy and multifocal neurological deficits which are difficult to distinguish from acute disseminated encephalomyelitis (ADEM). Current operational international consensus criteria for MS diagnosis in children with an ADEM-like first attack require confirmation by two or more non-ADEM like attacks, or one non-ADEM attack followed by accrual of clinically silent lesions. Although children with an ADEM-like first MS attack are more likely than children with monophasic ADEM to have one or more non-enhancing T1 hypointense lesions, two or more periventricular lesions, and the absence of a diffuse lesion pattern, these features are not absolutely discriminatory. Furthermore, MRI scans of children with monophasic ADEM typically demonstrate multiple variably enhancing lesions (often more than two) typically located in the juxtacortical white matter, infratentorial space, and spinal cord. For such patients serial clinical and MRI observations are required to confirm a diagnosis of MS. In this young age group, there can be marked lesion resolution following an initial attack prior to emergence over time of new lesions and attacks leading to a diagnosis of MS.

Diseases determining multifocal lesion of the CNS with recurrent episodes, such as systemic or brain vasculitis, Behçet's disease, Sjögren's syndrome and sarcoidosis, have to be considered in the differential diagnosis; in these cases visceral involvement and a positive result of immunological screening may be suggestive of an autoimmune systemic disease. Infective diseases such as neurosyphilis, AIDS, or Lyme's disease may present with multifocal involvement of the CNS with oligoclonal bands and intrathecal synthesis of IgG, but specific laboratory exams enable a differential diagnosis to be made. Some metabolic diseases with juvenile and adult onset, such as metachromatic leukodystrophy, globoid cell leukodystrophy, GM1 and GM2 gangliosidosis and adrenoleukodystrophy, may be in differential diagnosis with primary progressive MS. Finally, some intraspinal vascular malformations may mimic MS with a sub-acute onset and remission of symptoms.

Thus, if we consider alternative diagnoses apart from the examinations meeting the diagnostic criteria, the diagnostic protocol may be usefully integrated with further exams that, in agreement with clinical data, may include: erythrocyte sedimentation rate, c reactive protein, rheumatoid factor, anti-nuclear antibodies, anti-DNA antibodies, extractable nuclear antigen antibodies (for vasculitis), serum and CSF Venereal Disease Research Laboratory (for neurosyphilis), chest X-ray, Angiotensin I-converting enzyme assay (for sarcoidosis), serum and CSF anti-Borrelia burgdorferi antibodies, very long-chain fatty acid assay (for adrenoleukodystrophy), coagulation screening including antiphospholipid antibodies and protein C and S (for coagulopathy).

It is worth mentioning the differential diagnosis between multiple sclerosis and Devic's optic neuromyelitis (NMO) (see Table 6.1); in fact, there is increasing evidence of relapsing CNS demyelinating disease characterized by involvement of the optic nerves (unilateral or bilateral optic neuritis), severe myelopathy with MRI evidence of longitudinally extensive spinal cord lesions, often normal brain MRI (or with abnormalities atypical for MS), and serum aquaporin-4 (AQP4) autoantibodies [7]. It has been agreed that this phenotype should be separated from typical MS because of its different clinical course, prognosis, underlying pathophysiology and poor response to some of the available MS disease-modifying therapies. This disorder should be carefully considered in the differential diagnosis of all patients presenting clinical and MRI features that are strongly suggestive of NMO or NMO spectrum disorder, especially if (1) myelopathy is associated with MRI-detected spinal cord lesions longer than 3 spinal segments and primarily involving the central part of the spinal cord on axial sections; (2) optic neuritis is bilateral and severe or associated with a swollen optic nerve or chiasm lesion or an altitudinal scotoma; and (3) intractable hiccough or nausea/vomiting is present for more than 2 days, with evidence of a peri-aqueductal medullary lesion on MRI. In patients with such features, AQP4 serum testing should be used to help make a differential diagnosis between NMO and MS, to help in avoiding misdiagnosis, and to guide treatment.

Table 6.1 Differential diagnosis between MS and NMO. Modified from [10]

	Devic's disease	Multiple sclerosis
Distribution of symptoms and signs	Restricted to the optic nerves and spinal cord	Any white-matter tract
Attack severity	Usually severe	Usually mild
Head MRI	Usually normal/non-specific changes	Multiple periventricular white-matter lesions
Cord MRI	Longitudinally extensive central necrotic lesions	Multiple small peripheral lesions
CSF cells	Pleocytosis during attacks	Rarely > 25 white cells
Oligoclonal bands	Usually absent	Usually present
Permanent disability	Usually attack related	Usually in late progressive phase
Female patients	80–90 %	60–70 %
Coexisting autoimmunity	Frequent(30–40 %)	Less common
Serum neuromyelitis optica antibody	Present	Absent

The prognosis of MS is extremely variable and depends on several factors [8, 9]. Clinically negative factors are:

1. Male sex
2. Late onset
3. Short interval between first and second clinical attack
4. High frequency of relapses in the first 5 years of disease
5. Poli-symptomatic onset
6. Cerebellar, pyramidal, brainstem, urinary or psychiatric symptoms at the onset
7. Progressive course from the onset

MRI may be useful in making a prognostic evaluation. In particular, the number of lesions at the onset of a clinical symptom suggestive of MS is predictive for subsequent evolution to clinically definite MS, for the progression of disability (at least in the initial phase of the disease), and for the progression of lesion load . It has been estimated that the proportion of patients with an EDSS score of more than 3 after 10 years of disease belong to 75 % of the patients with more than 10 MRI lesions at the first brain MRI and only to 15 % of the group of patients with fewer than 10 lesions at the basal MRI. Finally, the Gd+ lesions correlate better with the number of relapses than with disability. Disability also correlates with lesion load in T1 and T2 weighted images, but especially with total and segmentary brain atrophy.

In conclusion, in the absence of symptoms or signs or specific laboratory tests, the diagnosis of MS remains exclusively clinical, but needs instrumental examinations to exclude other diseases which may have a similar remittent and exacerbating course. Moreover, in the absence of biomarkers predictive of clinical evolution, any prognostic evaluation requires a careful evaluation of clinical and MRI data.

References

1. Mcdonald WI, Compston A, Edan G et al (2001) Recommended diagnostic criteria for multiple sclerosis: guidelines from the International Panel on the Diagnosis of Multiple Sclerosis. Ann Neurol 50:121–127
2. Polman CH, Reingold SC, Edan G et al (2005) Diagnostic criteria for multiple sclerosis: 2005 revisions to the "McDonald Criteria". Ann Neurol 58:840–846
3. Polman CH, Reingold SC, Banwell B et al (2011) Diagnostic criteria for multiple sclerosis: 2010 revisions to the McDonald criteria. Ann Neurol 69:292–302
4. Barkhof F, Filippi M, Miller DH et al (1997) Comparison of MR imaging criteria at first presentation to predict conversion to clinically definite multiple sclerosis. Brain 120:2059–2069
5. Tintoré M, Rovira A, Martínez M et al (2000) Isolated demyelinating syndromes: comparison of different MR imaging criteria to predict conversion to clinically definite multiple sclerosis. Am J Neuroradiol 21:702–706
6. Andersson M, Alvarez-Cermeño J, Bernardi G et al (1994) Cerebrospinal fluid in the diagnosis of multiple sclerosis: a consensus report. J Neurol Neurosurg Psychiatry 57:897–902
7. Jarius S, Wildemann B (2010) AQP4 antibodies in neuromyelitis optica: diagnostic and pathogenetic relevance. Nat Rev Neurol 6:383–392
8. Myhr KM, Riise T, Vedeler C et al (2001) Disability and prognosis in multiple sclerosis: demographic and clinical variables important for the ability to walk and awarding of disability pension. Mult Scler 7:59–65
9. Confavreux C, Vukusic S, Adeleine P (2003) Early clinical predictors and progression of irreversible disability in multiple sclerosis: an amnesic process. Brain 126:770–782
10. Weinshenker BG (2003) Neuromyelitis optica: what it is and what it might be. Lancet 361:889–890

Assessment Instruments

7

Silvia Romano, Carlo Caltagirone, and Ugo Nocentini

In recent decades, also with regard to MS, rating scales have become essential tools both for assessing disability progression in clinical practice and as primary or secondary outcome measures in clinical trials.

The value of a scale is closely related to its clinical utility; a clinical scale should be short, easy to understand, quick to administer and clear to interpret. In addition, from a scientific point of view, it should have the following characteristics: reliability, validity and sensitivity.

Reliability concerns: the internal consistency of scale items, the reproducibility of scores when the scale is applied several times by the same operator (intra-operator) or by different operators (inter-operator), or to the same patient without modification of its state (test-retest).

There are many sources of error that may influence the scoring of a scale; the degree of reliability is indicative of the ability of the scale to rule out most errors.

In contrast, validity assesses if a scale is really able to measure what it is designed to measure. There are internal and external measures to quantify the validity of a scale: internal measures assess the scale scores and offer theoretical evidence that the variable has really been measured; external measures evaluate the correlations among the scores obtained with a new assessment tool and those obtained with similar instruments for which an acceptable validity has already been established. In this way it is possible to obtain empirical evidence that the tested variable really has been measured.

S. Romano (✉)
Centre for Experimental Neurological Therapies (CENTERS), Neurology Unit,
S. Andrea Hospital, University of Rome "La Sapienza", Rome, Italy
e-mail: silvia.romano@uniroma1.it

C. Caltagirone • U. Nocentini
Dipartimento di Neuroscienze, Università degli Studi di Roma "Tor Vergata", Rome, Italy
e-mail: c.caltagirone@hsantalucia.it; u.nocentini@hsantalucia.it

U. Nocentini et al. (eds.), *Neuropsychiatric Dysfunction in Multiple Sclerosis*,
DOI 10.1007/978-88-470-2676-6_7, © Springer-Verlag Italia 2012

The problem becomes even more complex when we consider that, in the case of MS, the aspects to be assessed are health status and disability, and both can be modified by several events (drug treatment or rehabilitation, but also any life event). In this perspective, therefore, the sensitivity of a scale must also be considered, meaning its capacity to detect even small variations of the measured parameter, preferably in a relatively short time.

Furthermore, the scale scores should present a normal distribution when applied to large populations, without "ceiling" or "floor" effect, making it possible to discriminate among patients with different degrees of disability.

In clinical practice, rating scales can be classified on the basis of the aspects evaluated: there are, in fact, illness-centered scales, which assess the severity of the disease in terms of symptoms and clinical signs, and patient-centered scales, which assess the physical, psychological and social impact of the disease from the patient's perspective (e.g., quality of life scales).

Regarding MS, several attempts have been made to create clinical scales which are able to assess the impact of the disease on patients; most clinicians and researchers have considered disability as the fundamental aspect to evaluate.

The *National Multiple Sclerosis Society (NMSS) Task Force on Outcome Measures*, held in Charleston in 1994, has established that the outcome measures in MS should (a) reflect the real level of severity and functional status of the patient, (b) be multidimensional, to identify the main aspects of the impact of the disease on the individual, (c) have scientific validity and (d) be able to detect changes over time [1].

The *Expanded Disability Status Scale (EDSS)* [2] is today still the scale most frequently used in studies and routine clinical evaluation to assess disability in MS. It was developed based on the *Kurtzke Disability Status Scale (DSS)* and is compiled by the neurologist in about 25 min. In both of these scales the scores of the different functional systems (FS), as described by Kurtzke, weigh on the general score and their application methods are identical.

The EDSS assesses the following functional systems: Pyramidal, Cerebellar, Brainstem, Sensory, Sphincteric, Visual, Mental or Cerebral and Other functions. For each function there is a score ranging from 0 to 5 or 0 to 6 points. The scores of each FS are not added because the EDSS provides a global score to identify the impairment according to deficits corresponding to different functional systems. Each score corresponds to ½ point, from normal neurologic function (EDSS 0.0) to death as a result of MS (EDSS 10).

Scores below 4.0 are established based on FS and relate to patients able to walk independently for at least 50 m; scores between 4.0 and 5.0 are based both on FS and walking ability, while scores between 5.5 and 8.0 are determined exclusively on the basis of walking ability and necessary aids.

However, this scale presents many problems: the EDSS is not a linear scale (the patients are distributed more frequently at some levels than others, with a kind of bimodal distribution at two focal points, scores 2–3 and 6), has limited credibility and low sensitivity, and walking impairment has a higher weight than other deficits (the upper range of the EDSS is not influenced by FS scores; fatigue, pain and mentation are not taken into due account).

Considering these limitations, several attempts have been made to produce new clinical scales such as the *Neurological Rating Scale (SCRIPPS)*, developed by Sipe in 1984 [3] and based on a standard neurological assessment plus the assessment of sphincter and sexual dysfunction; however, this scale has low validity and sensitivity data are not available. Other, less commonly used scales are the Troiano Functional Scale [4], the Illness Severity Scale [5], the Incapacity Status Scale and the Cambridge Basic MS Score (CAMBS) [6].

Of the new scales, the Multiple Sclerosis Functional Composite (MSFC), a clinical rating scale proposed in 1996 by a task force of the National Multiple Sclerosis Society (NMSS), plays an important role [7]. This task force, prompted by the poor reproducibility, low sensitivity to change and limited measurement properties of the previous scales used in MS, identified the principles and criteria for the creation of a new scale, leading to the development of the MSFC. This scale consists of three tests selected on the basis of data from longitudinal and natural history studies to assess clinically relevant variables. The three tests are: the nine-Hole Peg Test (9HPT) for upper limb functions, the Timed 25-ft (T-25f) Walk for ambulation and the 3 s-Paced Auditory Serial Addition Test (PASAT 3) for cognitive functions. Unlike the EDSS, the MSFC can also be administered by non-medical staff (technicians after appropriate training or paramedics), the time of administration is about 15 min and the required materials are easily available (a quiet room with a desk, and a corridor for the Timed 25-ft). The first test to be performed is the T-25f, which consists in measuring the time the patient takes to complete a 25-ft course as quickly as possible (but without running) twice (round trip). The 9HPT, the second test of the scale, assesses the time required for the patient to put nine pegs in the holes of a square board as quickly as possible and one at a time, and then remove them, again one at a time. The test is repeated four consecutive times, twice with the dominant hand and twice with the non-dominant hand. The PASAT, on the other hand, evaluates auditory information processing speed and flexibility, as well as calculation ability. Single digits are presented auditorily every 3 s: the patient must add each new digit to the one immediately prior to it. The test score is the total number of correct sums given (out of 60 possible) in each trial. The scores obtained in the various tests are combined to obtain a single score (Z-score), used to detect changes over time in a group of patients and compare them with each other. The Z-score is expressed in units of standard deviations, and indicates how much a patient's score is higher or lower than the mean score of a control population. The Z score is thus a continuous variable and more sensitive to clinical changes over 1-and 2-year intervals.

The MSFC is able to predict concurrent and subsequent EDSS changes [8], but compared to the latter scale has a greater correlation with MRI changes and with the degree of disability in daily life reported by patients [9].

The MSFC also has other advantages because, from a psychometric perspective, it has acceptable validity criteria, includes multidimensional assessments (considering many aspects of disability in MS patients), analyzes continuous variables

(Z-score) and uses standardized protocols of administration; moreover, in clinical trials, the possibility of administration by non-medical staff is a further guarantee that the physician-investigator will remain blind. However, it should be noted that some conditions different from those measured can affect test results (e.g., dysarthria affects PASAT performance, while visual dysfunctions influence the 9HPT score). Finally, the scale cannot be applied to patients in the later stages of the disease.

The most commonly used quality of life scales are the Multiple Sclerosis Impact Scale (MSIS-29) [10] and the Multiple Sclerosis Quality of Life (MSQoL)-54 [11].

The MSIS-29 is a self-administered test consisting of 29 questions, answered in a few minutes, that assess the patient's opinion of the impact of MS on daily life. Compared with other scales (EDSS and MSFC), the MSIS-29 is a valuable tool to assess and understand the physical and psychological impact of MS [12, 13].

The other scale, the MSQOL 54, developed by Vickrey and colleagues [11], is able to assess both general and MS-specific aspects relating to the quality of life. The scale has high test-retest reliability and high internal consistency and, since the central part consists of questions belonging to the 36-Item Short Form Health Survey, it is possible to compare data on the quality of life in MS patients with those in a general population. This tool has been used in a study on quality of care of MS patients, and has proved to be effective in evaluating the impact of management on the quality of life [14].

MS fatigue self-reported questionnaires are the most widely used method for measuring this symptom, in particular in clinical investigations. These scales are generally short, widely available and easy to administer. The most commonly used scale is the Fatigue Severity Scale (FSS) [15]. It consists of nine questions that assess the impact of fatigue on daily life. The patient is asked to choose a number from 1 to 7 that indicates how much she or he agrees with each statement. Another scale, the Modified Fatigue Impact Scale (MFIS) [16], has been recommended by the MS Council as an outcome measure for fatigue [17]. However, these scales do not seem to satisfy modern standards of outcome measurement [18, 19].

References

1. Rudick R, Antel J, Confavreux C et al (1997) Recommendations from the national multiple sclerosis society clinical outcomes assessment task force. Ann Neurol 42:379–382
2. Kurtzke JF (1983) Rating neurologic impairment in multiple sclerosis: an expanded disability status scale (EDSS). Neurology 33:1444–1452
3. Sipe JC, Knobler RL, Braheny SL, Rice GP, Panitch HS, Oldstone MB (1984) A neurologic rating scale (NRS) for use in multiple sclerosis. Neurology 34:1368–1372
4. Troiano R, Devereux C, Oleske J et al (1988) T cell subsets and disease progression after total lymphoid irradiation in chronic progressive multiple sclerosis. J Neurol Neurosurg Psychiatry 51:980–983
5. Mickey MR, Ellison GW, Myers LW (1984) An illness severity score for multiple sclerosis. Neurology 34:1343–1347
6. Mumford CJ, Compston A (1993) Problems with rating scales for multiple sclerosis: a novel approach – the CAMBS score. J Neurol 240:209–215

7. Cutter GR, Baier ML, Rudick RA et al (1999) Development of a multiple sclerosis functional composite as a clinical trial outcome measure. Brain 122:871–882

8. Fischer JS, Rudick RA, Cutter GR, Reingold SC (1999) The multiple sclerosis functional composite measure (msfc): an integrated approach to ms clinical outcome assessment. national ms society clinical outcomes assessment task force. Mult Scler 5:244–250

9. Rudick RA, Cutter G, Reingold S (2002) The multiple sclerosis functional composite: a new clinical outcome measure for multiple sderosis trials. Mult Scler 8:359–365

10. Hobart J, Lamping D, Fitzpatrick R, Riazi A, Thompson A (2001) The multiple sclerosis impact scale (msis-29): a new patient-based outcome measure. Brain 124:962–973

11. Vickrey BG, Hays RD, Harooni R, Myers LW, Ellison GW (1995) A health-related quality of life measure for multiple sclerosis. Qual Life Res 4:187–206

12. Hoogervorst EL, Zwemmer JN, Jelles B, Polman CH, Uitdehaag BM (2004) Multiple Sclerosis Impact Scale (MSIS-29): relation to established measures of impairment and disability. Mult Scler 10:569–574

13. Hobart JC, Riazi A, Lamping DL, Fitzpatrick R, Thompson AJ (2004) Improving the evaluation of therapeutic interventions in multiple sclerosis: development of a patient-based measure of outcome. Health Technol Assess 8:1–48

14. Solari A (2005) Role of health-related quality of life measures in the routine care of people with multiple sclerosis. Health Qual Life Outcomes 18:3–16

15. Krupp LB, LaRocca NG, Muir-Nash J, Steinberg AD (1989) The fatigue severity scale. Application to patients with multiple sclerosis and systemic lupus erythematosus. Arch Neurol 46:1121–1123

16. Fischer JS, LaRocca NG, Miller DM, Ritvo PG, Andrews H, Paty D (1999) Recent developments in the assessment of quality of life in multiple sclerosis (MS). Mult Scler 5:251–259

17. Panel of the Multiple Sclerosis Council for Clinical Practice Guidelines (1998) Fatigue and Multiple Sclerosis: Evidence-Based Management Strategies for Fatigue in Multiple Sclerosis. Paralyzed Veterans of America, Washington, DC

18. Mills R, Young C, Nicholas R, Pallant J, Tennant A (2009) Rasch analysis of the Fatigue Severity Scale in multiple sclerosis. Mult Scler 15:81–87

19. Mills RJ, Young CA, Pallant J, Tennant A (2009) Rasch analysis of the modified fatigue impact scale (mfis) in multiple sclerosis. J Neurol Neurosurg Psychiatry 81:1049–1051

Neuroimaging in Multiple Sclerosis

8

Gioacchino Tedeschi, Renato Docimo, Alvino Bisecco, and Antonio Gallo

8.1 Introduction

Magnetic Resonance Imaging (MRI) is the principal neuroimaging technique applied to Multiple Sclerosis (MS). Since the early 1990s, conventional MRI (c-MRI) has become a fundamental tool for MS diagnosis, management and research [1]. The distinctive features of c-MRI are represented by its high sensitivity to focal white matter (WM) lesions as well as infraclinical disease activity which is characterized by the appearance of new lesions in the absence of signs and/or symptoms of clinical relapse. In acknowledging these properties, MS diagnostic criteria have implemented specific c-MRI criteria, so that this technique has become the main paraclinical tool for supporting and reaching MS diagnosis [2–4]. More recently, non-conventional MRI (nc-MRI) techniques have allowed to investigate in vivo the pathophysiology of MS. By using nc-MRI, it has been possible to confirm, or even anticipate, results of neuropathological studies showing the presence of a diffuse microscopic damage outside focal WM lesions, involving both normal-appearing WM (NAWM) and gray matter (GM). nc-MRI-derived metrics have also prompted considerable interest since their stronger correlation with clinical disability scores than c-MRI-derived metrics. Finally, functional MRI (fMRI), the most recent and advanced nc-MRI technique, has made it possible to investigate mechanisms of cortical neuroplasticity in MS, providing very promising results.

G. Tedeschi (✉) • R. Docimo • A. Bisecco • A. Gallo
Istituto di Scienze Neurologiche, Seconda Università di Napoli, Naples, Italy
e-mail: gioacchino.tedeschi@unina2.it; renato.docimo@gmail.com; alvino.bisecco@gmail.com; antonio.gallo@unina2.it

U. Nocentini et al. (eds.), *Neuropsychiatric Dysfunction in Multiple Sclerosis*,
DOI 10.1007/978-88-470-2676-6_8, © Springer-Verlag Italia 2012

8.2 Conventional MRI (C-MRI)

8.2.1 Clinical Applications

To identify focal WM lesions in MS, c-MRI relies on T2/PD (Proton Density) and T1 sequences, the latter usually acquired pre- and post-contrast medium administration (gadolinium-DTPA [Gd] at the standard dosage of 0,1 mmol/Kg). Another commonly acquired sequence is the T2-FLAIR (Fluid Attenuated Inversion Recovery) which allows for a better visualization of periventricular and juxtacortical lesions due to suppression of the free water (i.e., Cerebro-Spinal Fluid – CSF) signal.

Compared with surrounding tissues, focal WM lesions appear hyperintense on T2, DP and T2-FLAIR images and isointense or hypointense on pre-contrast T1 images. Permanent T1 hypointense lesions are commonly defined as black holes (BHs). Finally, new and reactivated lesions appear hyperintense on post-contrast T1 sequences due to Gd uptake secondary to blood–brain barrier (BBB) damage.

Focal WM lesions are detectable in more than 95 % of patients with a definite diagnosis of MS [5]. WM lesions show multifocal distribution, with most of them located in the periventricular regions; the shape is typically ovoid/elongated, with the major axis perpendicular to the ventricles, due to the presence of inflammatory infiltrates along deep radial venules. The corpus callosum, infratentorial and subcortical/juxtacortical regions are frequently involved in MS as well. Moreover, optimized MRI sequences make it possible to identify optic nerve and spinal cord (SC) lesions.

Using standard T2/PD images, SC MRI has shown the presence of lesions in a high (47–90 %) percentage of MS patients. These data, coupled with those showing that SC lesions can be asymptomatic, prompt the acquisition of SC MRI in all cases with clinical findings suggestive of MS and a negative brain MRI [6]. SC MS lesions are usually characterized by: (1) preferential upper cervical cord localization (C1-C4); (2) extension no longer than two vertebral segments; (3) partial and peripheral cord involvement (on the transverse plane); (4) absence of SC swelling; and (5) absence of a T1-hypointense signal [7]. Finally, it is worth to note that recently developed sequences such as STIR (Short Tau Inversion Recovery) have shown a better sensitivity to MS SC lesions than T2/PD [8].

Some of the above-mentioned c-MRI features have been tested and incorporated into diagnostic criteria for MS (for more details see the chapter: MS diagnosis), making them more objective and sensitive [2–4]. Nevertheless, it should be recognized that c-MRI suffers from inherent limitations related not only to its non-quantitative nature, but also to its low pathological specificity. T2 images (the most widely-used c-MRI sequences in MS), indeed, do appear to be extremely sensitive to MS focal WM lesions, but perform poorly in distinguishing between lesions with different pathological substrates such as edema, inflammation, myelin and/or axonal loss, necrosis, etc. Therefore, independently of their prevalent pathology, MS focal WM lesions do appear hyperintense on T2 images. Needless to say, the above-mentioned limitations have a negative impact on differential diagnosis between MS and MS-like disorders (see, for example,

ADEM, Neuromyelitis Optica, CNS vasculitis, etc.), where the pattern of T2 hyperintensities can resemble MS, despite different pathological substrates [9, 10].

In partial contrast to T2 images, T1 post-contrast sequences have a higher pathological specificity, thus making it possible to recognize at least two types of MS WM lesions: (1) chronic and/or inactive lesions, which do not show Gd enhancement and appear as isointense or hypointense; (2) active – new or reactivated – enhancing lesions which uptake Gd and appear hyperintense.

As regards chronic hypointense lesions (i.e. BHs), neuropathology studies have shown that the degree of hypointensity is associated to a more severe tissue damage (i.e. axonal loss) inside WM lesions [11].

c-MRI has significantly contributed to the understanding of the temporal evolution of MS. In particular, c-MRI studies have consistently shown that radiological activity, as measured by serial MRIs, is much higher than clinical activity [1]. This means that new lesions can appear as well as pre-existing lesions can reactivate, enlarge or eventually disappear, even in the absence of any clinical signs/symptoms.

To summarize, c-MRI provides relevant information on MS dissemination in space and time, but also shows some limitations that justify the frequent clinical-radiological mismatch observed in MS patients [12–14]. These limitations are represented by: (1) low pathological specificity (as explained above); (2) inability to evaluate the microscopic brain tissue damage outside focal/macroscopic WM lesions, that certainly impact on clinical disability; and (3) inability to explore cortical functional changes (i.e.neuroplasticity) associated to MS.

Finally, it should be considered that the clinical-radiological mismatch also depends on the inadequacy of many clinical scales. In this regard it is worth noting that the Expanded Disability Status Scale (EDSS), the most widely used disability scale in MS, is: (1) not objective, (2) ordinal and not continuous; and (3) too biased toward motor/ambulation deficits [15].

8.2.2 Research Applications

c-MRI significantly contributes to MS clinical research in the following two areas: (1) natural history studies; (2) phase II/III clinical trials for the assessment of new therapies.

As regards natural history studies, c-MRI has made it possible to define the site and morphology, as well as the mode of appearance and temporal evolution, of focal WM lesions [1]. On this latter point, c-MRI has contributed to defining the characteristics of lesions – such as size and type/duration of contrast-enhancement – that are predictive of persistent and severe tissue damage, as reflected by chronic BHs formation [16]. Moreover, according to neuropathological data [17], c-MRI has supported the notion that the patterns of enhancing lesion evolution in MS are rather uniform within patients [18]. Natural history studies have therefore generated the knowledge and set up the methodology for the widespread application of c-MRI to clinical settings (diagnosis, follow-up, etc.) and clinical trials [19].

The first phase II/III clinical trial using c-MRI as a secondary outcome measure was performed in 1993. Evidence from c-MRI data supported the approval of the first formulation of Interferon-beta for relapsing-remitting MS (RRMS) [20]. Since then, all registered drugs for RRMS have supported their clinical efficacy with positive c-MRI results [21–24]. Moreover, all recent trials exploring the efficacy of available RRMS drugs in delaying the conversion of Clinically Isolated Syndrome (CIS) to clinically-definite MS (CDMS) have shown positive effects on c-MRI parameters [25–28].

c-MRI endpoints used in phase II/III clinical trials are usually based on serial (monthly to annual) MRI scans and are represented by: (1) number of active lesions, including new or enlarged lesions on T2/DP images and Gd-enhancing lesions on post-contrast T1 images; (2) volumes of T2-hyperintense and T1-hypointense lesions; (3) rate of conversion of new lesions into chronic BHs; (4) cross-sectional and longitudinal measures of brain and SC atrophy.

As regards atrophy measures, these have been introduced more recently, but are increasingly used in MS. Thanks to optimization of image acquisition (with the possibility of obtaining high-resolution images in a reasonable time) and analysis (with the development of accurate semi- or fully-automated tools) as well as good correlations with clinical disability scores, these measures have become a commonly-used surrogate marker for clinical studies.

Although c-MRI measures complement primary clinical endpoints such as relapse rate and EDSS, they still cannot substitute them in phase III clinical trials. The limitation of c-MRI to a secondary outcome arises from correlation studies showing an unsatisfactory relationship between c-MRI parameters and clinical measures. Moreover, it has been observed that while drugs reducing clinical activity usually determine a concomitant reduction of c-MRI activity, an effect on c-MRI activity is not invariably associated to a significant clinical effect.

To summarize, c-MRI is recommended in (1) explorative phase II studies assessing new drugs in definite forms of MS; (2) confirmative phase III studies in the same kind of MS patients; and (3) CIS studies assessing baseline risk and rate/time to conversion to definite MS.

In the first case (phase II studies), the monthly count of Gd-enhancing lesions is recommended as the primary efficacy endpoint [29, 30].

In the second case (phase III studies), the count/volume of T2, T1 and Gd-enhancing lesions as well as atrophy measures and the rate of conversion of new lesions to BHs [31] all represent secondary efficacy endpoints.

In the last case (CIS studies), c-MRI is used to select high risk patients, i.e. those presenting with typical demyelinating lesions, and evaluate the effect of the drug on the rate and time of conversion to definite MS according to MRI diagnostic criteria.

8.3   Non-Conventional MRI (Nc-MRI)

8.3.1   Introduction and General Techniques

nc-MRI techniques have been developed in recent years in order to overcome c-MRI limitations and clinical-radiologic mismatch observed in MS. The main advantages of these quantitative techniques are (1) ability to measure both the macroscopic tissue damage present at the level of focal WM lesion and the microscopic tissue damage known to occur outside focal lesions, and (2) possibility to study the mechanisms of cortical functional reorganization following MS tissue damage.

8.3.1.1 Magnetization Transfer Imaging (MTI)

Protons contained in animal tissues can be "free", when linked to water molecules (note: these protons are those generating most of the MRI signal), or "bound", when linked to large molecules such as myelin and axonal membranes in the nervous tissue [32]. Between the two pools of protons there is a continuous bidirectional exchange, so that if "bound" protons are selectively stimulated, part of their magnetization energy will transfer to "free" protons. By using dedicated acquisition and analysis tools, it is possible to compute the rate of this exchange voxel by voxel, obtaining quantitative Magnetization Transfer Ratio (MTR) maps. In these maps, regions with lower MTR values correspond to areas with a reduced exchange between the two pools of protons reflecting a lower content of macromolecules. MS pathology, a mainly represented by myelin and axonal loss, determines a significant drop of MTR values in affected areas [32].

Thanks to its sensitivity to tissue damage and its reliability and ease of application, MTI has become one of the most widespread nc-MRI techniques applied to MS.

Thus far MTI has been used – also in the context of multicenter studies – to characterize: (1) focal macroscopic WM lesions identified by c-MRI; (2) diffuse microscopic damage occurring at the level of NAWM (i.e. outside focal macroscopic WM lesions) and GM. With regard to the latter, it is worth noting that the study of NAWM and GM can be conducted either by a conventional region of interest (ROI) approach or a more comprehensive and unbiased approach based on histogram construction and analysis [33].

8.3.1.2 Diffusion Tensor Imaging (DTI)

In a fluid system, water molecules move randomly and the motion can be mathematically expressed by the diffusion coefficient. In biological systems instead, the intrinsic structure of the tissues (i.e. macromolecules, cell membranes, etc.) strongly restrains and directs the motion of water molecules, making the term "apparent diffusion coefficient" (ADC) more appropriate [34]. In particular, based on tissue architecture, water motion will be isotropic (i.e. similar in all directions), in the case of poorly-organized/not oriented tissues, or anisotropic (i.e. predominant in one direction), in the case of well-organized/oriented tissues.

If we consider brain tissues, GM and WM have distinct and peculiar micro-structural characteristics. In fact, while GM is a grossly homogeneous tissue allowing a substantially isotropic motion of water molecules, WM is a highly compact and oriented tissue, strongly favoring water motion along the major fiber axis rather than perpendicular to it.

Since a single coefficient (i.e. ADC) appears insufficient to fully describe diffusion in anisotropic tissues, the reconstruction of a tensor by means of DTI has allowed a better characterization of the architecture, geometry and orientation of anisotropic structures such as WM [35]. The two principal DTI-derived metrics are: (1) mean diffusivity (MD), which represents an index of global diffusion, and (2) fractional anisotropy (FA), a scalar index of anisotropy ranging from 0 (mini-mum/no anisotropy) to 1 (maximum anisotropy) [36]. Similarly to what we described for MTI, the processing of DTI data produces quantitative MD and FA maps where ROI- or histogram-based analyses can be run [37].

More recently, a refinement of DTI acquisition and data analysis has allowed the development of tractography, which can be used to reconstruct and finely analyze brain and spine WM fiber tracts [37].

MS pathology determines a focal and diffuse (microstructural) brain and SC damage, particularly at the level of WM. Such pathological changes usually lead to increased MD and reduced FA values. DTI and tractography are therefore particularly suited to make a fine in vivo characterization of MS tissue damage, particularly at the level of WM [38].

8.3.1.3 Proton Magnetic Resonance Spectroscopy (1H-MRS)

1H-MRS is a particularly useful technique to study in vivo biochemical alterations occurring in MS brains. [39]. 1H-MRS is based on the possibility of capturing the signal of protons bound to molecules other than water. This method relies on the physical principle that nuclei of the same atomic species resonate at different frequencies depending on their chemical environment. The spectrum obtained by 1H-MRS consists of a series of signal peaks whose height is proportional to the concentration of each peak-associated proton-containing metabolite.

In the study of MS, different 1H-MRS acquisition techniques can be used: (1) single-voxel 1H-MRS, which provides information on a single volume of variable size, positioned at the level of focal lesions or normal appearing tissue; advantages: ease of acquisition and data analysis as well as executability on most MRI scanners; disadvantages: impossibility of evaluating multiple brain regions in a diffuse/multifocal disease such as MS; (2) multivoxel 1H-MRS, which provides useful information on multiple voxels belonging to the same brain slice/slab; advantages: possibility of studying large portions of the brain; disadvantages: longer acquisition times and more complex data analysis than single-voxel 1H-MRS.

The principal brain metabolites detectable by 1H-MRS using long echo times (TE; 135–270 msec) are: (1) choline-containing compounds (Cho) and lipids, which are the main constituents of cell membranes and myelin; (2) creatine (Cr), a marker of cellular energy metabolism; (3) N-acetylaspartate (NAA), a metabolite almost exclusively present at the level of neurons and axons; (4) lactate (Lac), a product of anaerobic oxidative metabolism. Notably, most of the above-mentioned metabolites can be normalized to Cr – which is relatively stable throughout the brain – in order to have data that are more easily comparable between scans or subjects.

Other metabolites can be captured and analyzed when acquiring 1H-MRS with short TE (20–30 msec): myo-inositol, glutamate and glutamine.

In summary, 1H-MRS has proved particularly useful in the study of MS because of its complementary role in tissue characterization along with other nc-MRI techniques. By using this technique, in particular, it has been possible to assess the amount of axonal damage – reflected by a reduction of NAA – which is the pathological substrate more closely tied to permanent disability in MS [40, 41].

8.3.1.4 Functional MRI (fMRI)

fMRI is increasingly used in neurological diseases to investigate cortical reorganization mechanisms (i.e. neuroplasticity) taking place in the damaged brain. fMRI is based on microcirculatory changes following neuronal activation elicited by specific tasks or baseline (resting state) network activity. Such microcirculatory changes, indeed, translates into BOLD (blood oxygenation level-dependent) fMRI signal changes that mainly depend on: (1) blood flow/volume; (2) oxy-hemoglobin/de-oxy-hemoglobin ratio [42].

8.3.2 Techniques Dedicated to Gray Matter

Even if WM involvement remains the most characteristic hallmark of MS, another pathological feature of the disease is represented by the presence of a focal and diffuse GM damage, particularly evident at the level of cortical GM (cGM).

Unfortunately, visual detection of GM damage in vivo by c-MRI is strongly limited by all the following factors: (1) small size of GM abnormalities; (2) low contrast between affected and unaffected GM; (3) reduced inflammatory activity with very few contrast-enhancing lesions; (4) partial volume effects, mostly due to adjacent CSF, with detrimental effects on both cGM lesion detection and volume measurement; (5) unavailability in routine clinical practice of high-field MRI scanners (≥ 3.0 T) as well as MRI sequences calibrated for GM lesion detection.

The recent application and diffusion of nc-MRI techniques has changed the scenario, allowing to quantify GM damage and correlate it with clinical (physical and cognitive) disability [43, 44]. nc-MRI techniques, although mostly used for research purposes, have taken advantage of: (1) availability of high field MRI scanners (≥ 3 T) with multichannel coils; (2) development and optimization of GM-tailored sequences; (3) implementation of image processing analysis; (4) development of automatic and accurate GM segmentation techniques allowing the assessment of global and regional GM atrophy as well as cortical thickness (CTh) [45–47].

As regards MRI scanners, it has been clearly shown that high-field magnets (≥ 3.0 T) have a higher sensitivity to GM damage than standard ones (≤ 1.5 T) [48–51].

The recent development of optimized sequences has also greatly improved the in vivo detection and classification of cGM focal lesions (cGM-FLs) in MS. Among others, high-resolution MPRAGE, 2D/3D-FLAIR and Double Inversion Recovery (DIR) have all showed a consistently higher sensitivity to cGM-FLs than conventional T2/T1 images [52–55]. The DIR sequence, in particular, is able to distinguish pure cGM lesions (histologically classified as types II, III/IV) from mixed leucocortical GM lesions (type I) [48, 56, 57].

Among new image processing algorithms, those able to average multiple acquisitions of c-MRI sequences are worthy of mention. By using this approach, indeed, it is possible to increase the contrast to noise ratio and make it easier to identify cGM-FLs [58]. The main shortcomings of such an approach are represented by: (1) longer MRI sessions for multiple acquisitions; (2) methodological problems related to co-registration of multiple images, especially in presence of motion artifacts.

Automatic and accurate GM segmentation techniques have allowed to explore the global (focal and diffuse) burden of GM damage in MS and to test its relationship with clinical variables.

These techniques, that are able to measure both volume and thickness of cGM, rely on the precise allocation of each voxel to a different tissue class (e.g. GM, WM, cerebrospinal fluid). The quality of segmentation depends on the software used as well as the quality of MRI source images. The main requirements of these latter should be: (1) high spatial resolution; (2) good contrast between tissues (i.e. GM and WM); (3) homogeneous signal with few artifacts.

The most commonly used tools for segmentation and measurement of volume/thickness of cGM are: (1) *Structural Image Evaluation using Normalization of Atrophy*, for longitudinal (SIENA) and cross-sectional (SIENAX) volumetric studies; (2) *Voxel-Based Morphometry* (VBM), which is implemented in Statistical Parametric Mapping (SPM software package) and allows the investigation of regional GM atrophy [47]; and (3) FreeSurfer, which measures GM volumes as well as whole and regional cGM thickness [46, 59–63].

8.4 Non-conventional MRI: Its Contribution to the Study of White Matter

8.4.1 Focal White Matter Lesions

8.4.1.1 MTI

There is much evidence to suggest that MTR values measured at the level of focal WM lesions are inversely correlated with the degree of tissue damage. In particular, a reduction of MTR correlates with: (1) axonal loss [64], as measured by neuropathology; (2) NAA reduction [65], as measured by 1H-MRS; (3) degree of BHs hypointensity, as measured on T1 images [64, 66]. Interestingly, a reduction of MTR has also been found at the level of focal lesions weeks to months before their appearance on c-MRI sequences [67–70]. After lesion appearance MTR values tend to slowly recover, reaching normal baseline values in around 50% of the lesions. All the above-mentioned MTR changes are thought to reflect axonal loss and demyelination/remyelination processes more than edema/inflammation evolution.

In conclusion, MTI is a quantitative MRI technique that is able to characterize and monitor MS focal lesions and as such has a great potential for testing the protective/reparative effect of experimental therapies in MS.

8.4.1.2 DTI

MS focal lesions show reduced FA and increased MD values when compared to the NAWM of MS patients and the WM of healthy controls (HCs) [71]. These data suggest an expansion of the extracellular space and a loss of tissue anisotropy, both secondary to the damage of (microscopic) biological barriers (e.g. myelin, axon, etc.) physiologically restricting/directing water diffusion. In particular, while axonal loss and demyelination are both associated with an increase of MD and a decrease of FA, an intense reactive gliosis seems to be associated to a reduction of both FA and MD. Differently, the impact of inflammation on water diffusion is more variable and unpredictable and depends on the prevalence of edema, cellular infiltrates or degradation products.

8.4.1.3 1H-MRS

Using 1H-MRS, acute MS lesions show an increase of Cho, Lac, lipids and myo-inositol that essentially reflects myelin damage [72, 73]. All these changes are usually associated with a variable NAA reduction related to axonal damage/dysfunction [74]. Following the acute phase, lesion metabolites tend to gradually normalize, with the exception of NAA, which recovers to baseline values only in a few cases [75]. NAA is drastically and irreversibly reduced in progressive forms of MS, but is relatively preserved in benign forms of MS, thus reinforcing the clinical relevance of NAA levels [76–78].

8.4.2 Normal Appearing White Matter

8.4.2.1 MTI

Post-mortem studies have consistently showed diffuse pathological changes in the NAWM of MS patients. The substrates of such abnormalities include edema, gliosis, perivascular inflammation, demyelination and axonal loss [79, 80]. MTI is able to detect and quantify this microscopic damage [81], particularly in regions of the NAWM where focal lesions will subsequently arise [68, 82]. Using a histogram analysis approach, NAWM MTR values have been found more reduced in progressive forms of MS than in relapsing-remitting and benign forms of the disease [83–89]. These results suggest that NAWM damage increases with disease duration and severity. Accordingly, studies exploring the relationship between histogram-derived MTR values and clinical (physical and cognitive) disability scores found significant correlations [90, 91].

8.4.2.2 DTI

Similarly to MTI, DTI has been shown to be sensitive to microscopic damage in MS. In particular, increased MD and decreased FA values have been repeatedly detected in the NAWM of MS patients when compared to those measured in the WM of HCs [92, 93]. Similiar results have been obtained in CIS patients, confirming the extreme sensitivity of this technique [94, 95]. Interestingly, no correlations have been found between DTI parameters and T2-lesion load [37], thus suggesting that diffuse NAWM damage does not simply reflect wallerian degeneration associated to axonal transection occurring inside focal MS lesions.

8.4.2.3 1H-MRS

Significant NAA reductions have been repeatedly reported in the NAWM of CDMS patients [96, 97] and are related to physical disability [98]. These findings might reflect: (1) wallerian degeneration secondary to axonal transection involving fibers passing through focal WM lesions [79]; (2) NAWM changes preceding lesion appearance (as demonstrated by MTI [68] and DTI [99] studies); (3) axonal metabolic dysfunction secondary, for example, to mitochondrial failure [100]. Interestingly, transient/reversible reductions of NAA have also been reported, particularly in CIS and early MS [73, 101, 102]. More recently, an increase of myo-inositol has been detected in the NAWM of CIS patients (particularly in those converting to CDMS), thus suggesting a role for this metabolite as an early marker of NAWM microscopic damage [103, 104]. In the paper from *Wattjes et al.*, in particular, the authors hypothesize that specific metabolic changes occurring at the time of the first clinical event might be an important prognostic marker for the future clinical course [104].

8.5   Non-conventional MRI: Its Contribution to the Study of Gray Matter

8.5.1   Gray Matter Focal Lesions

cGM-FLs are detectable since the early phases of MS and rapidly increase in the chronic progressive stages of the disease [57]. Accordingly, the lowest rate of cGM-FL appearance has been detected in benign forms of MS [105]. The rising interest in cGM-FLs is justified by the increasing availability of new MRI sequences/scanners (that improve their visualization) as well as the results of transverse and longitudinal studies showing significant associations between volume/number of cGM-FLs and cognitive/physical disability scores [106–109]. Further supporting these latter evidences, the number/volume of cGM-FLs has been found to strongly correlate with cGM atrophy, a well-recognized surrogate marker of clinical disability [107–109].

8.5.2   Global Gray Matter: MTI, DTI, 1H-MRS Data

Several nc-MRI studies have confirmed in vivo pathological data reporting a diffuse involvement of GM in MS [43, 44, 110]. A number of investigations have indeed shown MTR and NAA reductions as well as significant MD/FA changes in the GM, particularly at the level of cGM [43, 44, 110]. Notably, the above-mentioned changes were more frequent and pronounced in advanced stages of the disease and correlated, in most cases, with clinical disability. Altogether these data seem to suggest that a diffuse GM damage might play a crucial role in determining irreversible disability in MS [43, 44, 110].

8.5.2.1 MTI
Lower MTR values are detectable at the level of cGM from the earliest stages of MS and become more pronounced during late (progressive) stages of the disease [90, 111]. Contrariwise, MTR data on deep GM (dGM) are few and not consistent [112]. Finally, a series of transverse and longitudinal studies have shown significant correlations between GM MTR values and physical [111]/cognitive [113, 114] disability as well as medium to long term disease evolution [115–117].

8.5.2.2 DTI
Several studies have found abnormal DTI-derived measures (MD and FA) in both the cGM and dGM of MS patients [43, 44, 93, 110, 118–121]. Such abnormalities were more marked in chronic progressive phases of the disease [118, 119]. Interestingly, GM MD values showed to be sensitive to disease evolution [122, 123] and good predictors of clinical [124] and cognitive status [125, 126].

8.5.2.3 1H-MRS
RRMS and SPMS patients have consistently shown a reduction of NAA levels in the cGM and dGM when compared to HCs [127–131]. Notably, in one study a peak related to myelin degradation products was also detected at the level of cGM [130]. More recently, a new 1H-MRS methodology, i.e. whole brain NAA (WBNAA), has been developed to measure the absolute content of NAA in the entire brain [132, 133]. By using this technique a significant reduction of WBNAA was found in all MS phenotypes and particularly in patients with advanced/progressive disease [134–136]. Interestingly, these studies did not report significant correlations between WBNAA and number/volume of focal WM lesions, again confirming the existence of a mismatch between diffuse/microscopic and focal/macroscopic damage in MS.

8.5.3   Global Gray Matter: Volumetric and Thickness Data

Global atrophy measures have documented a significant brain and SC shrinkage from the earliest stages of MS [44]. These measures have raised considerable interest in the MS community since their stronger correlation with clinical disability scores than focal WM lesions [137–140]. Moreover, atrophy measurement tools, thanks to continuous technical improvements, have become so sensitive to MS-related atrophy that 6–12 months can be sufficient to capture minimal but already significant volume losses [141–145].

Cognitive impairment (CI) is very common in MS patients [146]. Studies exploring the association between GM atrophy and CI have produced very promising results. The Florence group, in particular, first reported the presence of significant cGM atrophy in MS patients with CI and successively a higher rate of cGM loss in patients with cognitive performances worsening during a 2.5-year follow-up [143, 144]. Notably, these results have been subsequently confirmed and expanded by other research groups [107, 145, 147]. Only recently, in addition, atrophy of dGM structures such as thalamus, caudate and hippocampus has been recognized as another relevant factor in determining CI in MS [148–150].

Fatigue is another very common and disabling symptom of MS whose pathophysiology is largely unknown [151]. Interestingly, highly-fatigued MS patients have showed lower GM and WM volumes than HCs and non-fatigued patients [152].

Finally, mood disorders – frequently observed in MS – have also been related to GM atrophy, particularly at the level of fronto-temporal regions [153–155].

Moving from global to regional GM atrophy, it is mandatory to mention the VBM approach, which allows to compare average GM maps of different groups of subjects in order to extract areas/regions where GM is significantly reduced/increased compared to a control group. By using VBM, several studies have shown a significant GM loss at the level of fronto-temporo-parietal regions as well as the dGM and cerebellum of MS patients [155]. Notably, GM distribution has been linked to disease phenotype [156] and CI [157]. Finally, VBM studies have also shown the selective involvement of fronto-parietal circuits in fatigued MS patients [158, 159].

Another possible approach to the study of the cGM in MS is the measurement of CTh. CTh has been found to be globally reduced in MS and to correlate with both disability (physical and cognitive) and fatigue scores [147, 160–162]. *Calabrese* et al. [163], in particular, reported a significant cortical thinning since the earliest stages of the disease, as well as a significant relationship between clinical symptoms and regional CTh. Finally, a recent study by *Pellicano* et al. [164] showed an association between fatigue and cortical thinning of the posterior parietal lobe, thus suggesting that this symptom may be related to dysfunction of cortical areas deputed to motor planning and sensory integration.

In conclusion, the in vivo MRI assessment of GM in MS has not only confirmed and expanded neuropathologic data but has also provided a series of new potential surrogate markers for clinical trials.

8.5.4 Global Gray Matter: Functional MRI Data

Clinical MS symptoms and signs are the result of a complex interplay between tissue damage, tissue repair and cortical plasticity.

fMRI is the most commonly used tool for examining cortical reorganization – also known as neuroplasticity – and its role in functional recovery in MS and other neurologic diseases [165, 166].

As a general rule, MS patients performing a specific (motor, sensitive, cognitive, etc.) task properly show an increased activation of the cortical network normally deputed to that task (as observed in HCs). Differently, MS patients with longer disease duration and/or higher clinical disability, tend to expand the cortical network, thus recruiting additional secondary (task-related) areas to maintain a normal function/behaviour [164, 165]. Later on, when functional/behavioural output itself is compromised, brain cortical activations appear sparse and aimless, if not detrimental [164–166].

Of particular interest, fMRI changes detected in MS patients have been frequently associated with the extent and severity of focal and diffuse brain tissue damage, as measured by c- and nc-MRI.

More recently, fMRI studies conducted in resting conditions (i.e. without any stimulus or task) have allowed to investigate several functionally-relevant resting-state networks (RSNs) normally active in the healthy brain performing no task [167, 168]. One of the major advantages of the resting-state fMRI (RS-fMRI) approach is the avoidance of the well-known task-related fMRI limitations [167]. One of the most robust and reliable RSNs is the so-called default-mode network (DMN), which encompasses the medial prefrontal, anterior cingulate, posterior cingulate/precuneus, and lateral parietal cortices [169]. The DMN was initially identified because of its reliable and consistent deactivation during cognitive tasks, leading to the hypothesis that this network was involved in cognitive postprocessing at rest [170]; in support of this speculation was the finding that DMN is completely silenced in comatose patients [171].

The few studies conducted so far in MS have shown that CI is associated to a functional disconnection of the anterior component of DMN (i.e. anterior cingulate cortex) [172, 173]. In another study conducted on CIS patients an increased synchronization was found in several RSNs, including the DMN and the sensori-motor network. Interestingly, these changes were not observed in RRMS patients bearing a more severe structural damage [174], thus suggesting that plasticity of the RSNs might be an early and limited phenomenon.

To summarize, a minimum amount of structural damage seems enough to trigger significant neuroplastic changes in MS brains. In order to preserve function/behaviour an increased recruitment of task-related cortical areas as well as rein-forcement of brain RSNs connectivity might be the first steps in cortical adaptive reorganization. As the disease progresses to a more advanced phase – with structural damage that builds up – brain neuroplastic changes proportionally increase/expand in order to limit clinical worsening. At a certain some point, however, structural brain damage reaches a level such that an adaptive/compensatory functional reorganization can no longer take place [175, 176].

Before concluding, it is worth remembering that interpretation of fMRI data always requires caution, since brain functional reorganization is an extremely dynamic and variable phenomenon. Finally, the comparison of different fMRI studies is complicated by differences in: (1) subjects selection criteria, and (2) task/resting state fMRI design/protocols.

References

1. Miller DH, Grossman RI, Reingold SC, McFarland HF (1998) The role of magnetic resonance techniques in understanding and managing multiple sclerosis. Brain 1(121):3–24
2. McDonald WI, Compston A, Edan G et al (2001) Recommended diagnostic criteria for multiple sclerosis: guidelines from the international panel on the diagnosis of multiple sclerosis. Ann Neurol 50:121–127
3. Polman CH, Reingold SC, Edan G et al (2005) Diagnostic criteria for multiple sclerosis: 2005 revisions to the "McDonald Criteria". Ann Neurol 58:840–846
4. Polman CH, Reingold SC, Banwell B et al (2011) Diagnostic criteria for multiple sclerosis: 2010 revisions to the McDonald criteria. Ann Neurol 69(2):292–302
5. Ormerod IEC, Miller DH, McDonald WI et al (1987) The role of NMR imaging in the assessment of multiple sclerosis and isolated neurological lesions: a quantitative study. Brain 110:1579–1616
6. Thorpe JW, Kidd D, Moseley IF et al (1996) Spinal MRI in patients with suspected multiple sclerosis and negative brain MRI. Brain 119:709–714
7. Gass A, Filippi M, Rodegher ME et al (1998) Characteristics of chronic MS lesions in the cerebrum, brainstem, spinal cord, and optic nerve on T1-weighted MRI. Neurology 50:548–550
8. Rocca MA, Mastronardo G, Horsfield MA et al (1999) Comparison of three MR sequences for the detection of cervical cord lesions in patients with multiple sclerosis. AJNR Am J Neuroradiol 20(9):1710–1716
9. Triulzi F, Scotti G (1998) Differential diagnosis of multiple sclerosis: contribution of magnetic resonance techniques. J Neurol Neurosurg Psychiatry 64(Suppl 1):S6–S14
10. Pittock SJ, Lucchinetti CF (2007) The pathology of MS: new insights and potential clinical applications. Neurologist 13(2):45–56, Review
11. van Walderveen MA, Kamphorst W, Scheltens P et al (1998) Histopathologic correlate of hypointense lesions on T1-weighted spin-echo MRI in multiple sclerosis. Neurology 50:1282–1288
12. Filippi M, Paty DW, Kappos L et al (1995) Correlations between changes in disability and T2-weighted brain MRI activity in multiple sclerosis: A follow-up study. Neurology 45:255–260
13. Kappos L, Moeri D, Radue EW et al (1999) Predictive value of gadolinium-enhanced MRI for relapse rate and changes in disability/impairment in multiple sclerosis: a metaanalysis. Lancet 353:964–969
14. Barkhof F (2002) The clinico-radiological paradox in multiple sclerosis revisited. Curr Opin Neurol 15(3):239–245, Review
15. Kurtzke JF (1983) Rating neurologic impairment in multiple sclerosis: an expanded disability status scale (EDSS). Neurology 33(11):1444–1452
16. Bagnato F, Evangelou IE, Gallo A et al (2007) The effect of interferon-beta on black holes in patients with multiple sclerosis. Expert Opin Biol Ther 7(7):1079–1091, Review
17. Lucchinetti C, Bruck W, Parisi J et al (2000) Heterogeneity of multiple sclerosis lesions: implications for the pathogenesis of demyelination. Ann Neurol 47:707–717
18. Minneboo A, Uitdehaag BM, Ader HJ, Barkhof F, Polman CH, Castelijns JA (2005) Patterns of enhancing lesion evolution in multiple sclerosis are uniform within patients. Neurology 65(1):56–61
19. Li DK, Li MJ, Traboulsee A, Zhao G et al (2006) The use of MRI as an outcome measure in clinical trials. Adv Neurol 98:203–226, Review
20. Paty DW, Li DK (1993) Interferon beta-1b is effective in relapsing-remitting multiple sclerosis. II. MRI analysis results of a multicenter, randomized, double-blind, placebo-controlled trial. UBC MS/MRI study group and the ifnb multiple sclerosis study group. Neurology 43(4):662–667

21. Simon JH, Jacobs LD, Campion M et al (1998) Magnetic resonance studies of intramuscular interferon beta-1a for relapsing multiple sclerosis. Ann Neurol 43:79–87
22. Li DKB, Paty DW, The UBC MS/MRI Analysis Research Group, PRISMS Study Group (1999) Magnetic resonance imaging results of the PRISMS trial: a randomized, double-blind, placebo-controlled study of Interferon beta-1a in relapsing-remitting multiple sclerosis. Ann Neurol 46:197–206
23. Comi G, Filippi M, Wolinsky JS, The European/Canadian Glatiramer Acetate Study Group (2001) European/Canadian multicenter, double blind, randomized, placebo-controlled study of the effects of glatiramer acetate on magnetic resonance imaging–measured disease activity and burden in patients with relapsing multiple sclerosis. Ann Neurol 49:290–297
24. Polman CH, O'Connor PW, Havrdova E et al (2006) A randomized, placebo-controlled trial of natalizumab for relapsing multiple sclerosis. N Engl J Med 354(9):899–910
25. Jacobs LD, Beck RW, Simon JH et al (2000) Intramuscular interferon beta-1a therapy initiated during a first demyelinating event in multiple sclerosis. N Engl J Med 343:898–904
26. Comi G, Filippi M, Barkhof F et al (2001) Early treatment of Multiple Sclerosis Study Group. Effect of early interferon treatment on conversion to definite multiple sclerosis: a randomised study. Lancet 357:1576–1582
27. Kappos L, Polman CH, Freedman MS et al (2006) Treatment with interferon beta-1b delays conversion to clinically definite and McDonald MS in patients with clinically isolated syndromes. Neurology 67:1242–1249
28. Comi G, Martinelli V, Rodegher M, et al (2009) PreCISe study group. Effect of glatiramer acetate on conversion to clinically definite multiple sclerosis in patients with clinically isolated syndrome (PreCISe study): A randomised, double-blind, placebo-controlled trial. Lancet 374(9700):1503–1511. Erratum. In: Lancet (2010), 375(9724):1436
29. Hauser SL, Waubant E, Arnold DL et al (2008) B-cell depletion with rituximab in relapsing-remitting multiple sclerosis. N Engl J Med 358(7):676–688
30. Bakshi R, Hutton GJ, Miller JR, Radue EW (2004) The use of magnetic resonance imaging in the diagnosis and long-term management of multiple sclerosis. Neurology 63(11 Suppl 5): S3–S11
31. Barkhof F, van Waesberghe JH, Filippi M et al (2001) European Study Group on Interferon beta-1b in secondary progressive multiple sclerosis. T(1) hypointense lesions in secondary progressive multiple sclerosis: effect of interferon beta-1b treatment. Brain 124 (Pt 7):1396–1402
32. McGowan JC (1999) The physical basis of magnetization transfer imaging. Neurology 53(5 Suppl 3):S3–S7, Review
33. van Buchem MA, McGowan JC, Grossman RI (1999) Magnetization transfer histogram methodology: its clinical and neuropsychological correlates. Neurology 53(5 Suppl 3): S23–S28
34. Le Bihan D, Breton E, Lallemand D et al (1986) MR imaging of intravoxel incoherent motions: application to diffusion and perfusion in neurologic disorders. Radiology 161:401–407
35. Basser PJ, Mattiello J, LeBihan D (1994) Estimation of the effective self-diffusion tensor from the NMR spin-echo. J Magn Reson B 103:247–254
36. Pierpaoli C, Jezzard P, Basser PJ, Barnett A et al (1996) Diffusion tensor MR imaging of the human brain. Radiology 201:637–648
37. Cercignani M, Inglese M, Pagani E, Comi G et al (2001) Mean diffusivity and fractional anisotropy histograms in patients with multiple sclerosis. AJNR Am J Neuroradiol 22:952–958
38. Ciccarelli O, Catani M, Johansen-Berg H, Clark C, Thompson A (2008) Diffusion-based tractography in neurological disorders: concepts, applications, and future developments. Lancet Neurol 7(8):715–727, Review
39. Sajja BR, Wolinsky JS, Narayana PA (2009) Proton magnetic resonance spectroscopy in multiple sclerosis. Neuroimaging Clin N Am 19:45–58, Review

40. Arnold DL, De Stefano N, Narayanan S, Matthews PM (2001) Axonal injury and disability in multiple sclerosis: Magnetic resonance spectroscopy as a measure of dynamic pathological change in white matter. In: Magnetic resonance spectroscopy in multiple sclerosis, Milan, Springer, pp 61–67

41. Sarchielli P, Presciutti O, Pelliccioli GP et al (1999) Absolute quantification of brain metabolites by proton magnetic resonance spectroscopy in normal-appearing white matter of multiple sclerosis patients. Brain 122:513–521

42. Ogawa S, Menon RS, Kim SG, Ugurbil K (1998) On the characteristics of functional magnetic resonance imaging of the brain. Annu Rev Biophys Biomol Struct 27:447–474

43. Geurts JJ, Barkhof F (2008) Grey matter pathology in multiple sclerosis. Lancet Neurol 7(9):841–851, Review

44. Pirko I, Lucchinetti CF, Sriram S, Bakshi R (2007) Gray matter involvement in multiple sclerosis. Neurology 68(9):634–642, Review

45. Nakamura K, Fisher E (2009) Segmentation of brain magnetic resonance images for measurement of gray matter atrophy in multiple sclerosis patients. Neuroimage 44(3):769–776

46. Fischl B, Dale AM (2000) Measuring the thickness of the human cerebral cortex from magnetic resonance images. Proc Natl Acad Sci 97(20):11050–11055

47. Ashburner J, Friston KJ (2000) Voxel-based morphometry – the methods. Neuroimage 11:805–821, Review

48. Wattjes MP, Lutterbey GG, Gieseke J et al (2007) Double inversion recovery brain imaging at 3 T: diagnostic value in the detection of multiple sclerosis lesions. AJNR Am J Neuroradiol 28(1):54–59

49. Geurts JJ, Blezer EL, Vrenken H et al (2008) Does high-field MR imaging improve cortical lesion detection in multiple sclerosis? J Neurol 255(2):183–191

50. Mainero C, Benner T, Radding A et al (2009) In vivo imaging of cortical pathology in multiple sclerosis using ultra-high field MRI. Neurology 73(12):941–948

51. Schmierer K, Parkes HG, So PW et al (2010) High field (9.4 Tesla) magnetic resonance imaging of cortical grey matter lesions in multiple sclerosis. Brain 133(Pt 3):858–867

52. Nelson F, Poonawalla A, Hou P et al (2008) 3D MPRAGE improves classification of cortical lesions in multiple sclerosis. Mult Scler 14(9):1214–1219

53. Tubridy N, Barker GJ, Macmanus DG (1998) Three-dimensional fast fluid attenuated inversion recovery (3D fast FLAIR): a new MRI sequence which increases the detectable cerebral lesion load in multiple sclerosis. Br J Radiol 71(848):840–845

54. Lazeron RH, Langdon DW, Filippi M et al (2000) Neuropsychological impairment in multiple sclerosis patients: the role of (juxta)cortical lesion on FLAIR. Mult Scler 6(4):280–285

55. Bakshi R, Ariyaratana S, Benedict RH, Jacobs L (2001) Fluid-attenuated inversion recovery magnetic resonance imaging detects cortical and juxtacortical multiple sclerosis lesions. Arch Neurol 58(5):742–748

56. Geurts JJ, Pouwels PJ, Uitdehaag BM et al (2005) Intracortical lesions in multiple sclerosis: improved detection with 3D double inversion-recovery MR imaging. Radiology 236(1):254–260

57. Calabrese M, De Stefano N, Atzori M et al (2007) Detection of cortical inflammatory lesions by double inversion recovery magnetic resonance imaging in patients with multiple sclerosis. Arch Neurol 64(10):1416–1422

58. Bagnato F, Butman JA, Gupta S et al (2006) In vivo detection of cortical plaques by MR imaging in patients with multiple sclerosis. AJNR Am J Neuroradiol 27:2161–2167

59. Dale AM, Fischl B, Sereno MI (1999) Cortical surface-based analysis. I. Segmentation and surface reconstruction. Neuroimage 9(2):179–194

60. Dale AM, Fischl B, Sereno MI (1999) Cortical surface-based analysis. II: Inflation, flattening, and a surface-based coordinate system. Neuroimage 9(2):195–207

61. Fischl B, van der Kouwe A, Destrieux C et al (2004) Automatically parcellating the human cerebral cortex. Cereb Cortex 14(1):11–22

62. Desikan RS, Ségonne F, Fischl B et al (2006) An automated labeling system for subdividing the human cerebral cortex on MRI scans into gyral based regions of interest. Neuroimage 31(3):968–980
63. Fischl B, Salat DH, Busa E et al (2002) Whole brain segmentation: automated labeling of neuroanatomical structures in the human brain. Neuron 33(3):341–355
64. Van Waesberghe JH, Kamphorst W, DeGroot CJ et al (1999) Axonal loss in multiple sclerosis lesions: magnetic resonance imaging insights into substrates of disability. Ann Neurol 46:747–754
65. Kimura H, Grossman RI, Lenkinski RE et al (1996) Proton MR spectroscopy and magnetization transfer ratio in multiple sclerosis: correlative findings of active versus irreversible plaque disease. AJNR Am J Neuroradiol 17:1539–1547
66. Loevner LA, Grossman RI, McGowan JC, Ramer KN et al (1995) Characterization of multiple sclerosis plaques with T1-weighted MR and quantitative magnetization transfer. AJNR Am J Neuroradiol 16:1473–1479
67. Dousset V, Gayou A, Brochet B, Caille JM (1998) Early structural changes in acute MS lesions assessed by serial magnetization transfer studies. Neurology 51:1150–1155
68. Filippi M, Rocca MA, Martino G, Horsfield MA et al (1998) Magnetization transfer changes in the normal appearing white matter precede the appearance of enhancing lesions in patients with multiple sclerosis. Ann Neurol 43:809–814
69. Goodkin DE, Rooney WD, Sloan R, Bacchetti P et al (1998) A serial study of new MS lesions and the white matter from which they arise. Neurology 51:1689–1697
70. Rocca MA, Mastronardo G, Rodegher M et al (1999) Long-term changes of magnetization transfer-derived measures from patients with relapsing-remitting and secondary progressive multiple sclerosis. AJNR Am J Neuroradiol 20:821
71. Werring DJ, Clark CA, Barker GJ, Thompson AJ et al (1999) Diffusion tensor imaging of lesions and normal-appearing white matter in multiple sclerosis. Neurology 52:1626–1632
72. Davie CA, Hawkins CP, Barker GJ, Brennan A et al (1994) Serial proton magnetic resonance spectroscopy in acute multiple sclerosis lesions. Brain 117:49–58
73. Narayana PA, Doyle TJ, Lai D, Wolinsky JS (1998) Serial proton resonance spectroscopic imaging, contrast-enhanced magnetic resonance imaging, and quantitative lesion volumetry in multiple sclerosis. Ann Neurol 43:56–71
74. De Stefano N, Matthews PM, Antel JP, Preul M, Francis G et al (1996) Chemical pathology of acute demyelinating lesions and its correlation with disability. Ann Neurol 38:901–909
75. De Stefano N, Matthews PM, Arnold DL (1995) Reversible decreases in N-acetylaspartate after acute brain injury. Magn Reson Med 34:721–727
76. Fu L, Matthews PM, De Stefano N, Worsley KJ et al (1998) Imaging axonal damage of normal-appearing white matter in multiple sclerosis. Brain 121:159–166
77. Arnold DL, Matthews PM, Francis GS, O'Connor J et al (1992) Proton magnetic resonance spectroscopic imaging for metabolic characterization of demyelinating plaques. Ann Neurol 31:235–241
78. Falini A, Calabrese G, Filippi M, Origgi D et al (1998) Benign versus secondary-progressive multiple sclerosis: the potential role of proton MR spectroscopy in defining the nature of disability. AJNR Am J Neuroradiol 19:223–229
79. Trapp BD, Peterson J, Ransohoff RM, Rudick R et al (1998) Axonal transection in the lesions of multiple sclerosis. New Engl J Med 338:278–285
80. Allen IV, McKeown SR (1979) A histological, histochemical and biochemical study of the macroscopically normal white matter in multiple sclerosis. J Neurol Sci 41:81–91
81. Filippi M, Campi A, Dousset V et al (1995) A magnetization transfer imaging study of normal appearing white matter in multiple sclerosis. Neurology 45:478–482
82. Pike GB, De Stefano N, Narayanan S, Worsley KJ et al (2000) Multiple sclerosis: magnetization transfer MR imaging of white matter before lesion appearance on T2-weighted images. Radiology 215:824–830

83. Filippi M, Iannucci G, Tortorella C, Minicucci L et al (1999) Comparison of MS clinical phenotypes using conventional and magnetization transfer MRI. Neurology 52:588–594

84. Rovaris M, Bozzali M, Santuccio G et al (2000) Relative contribution of brain and spine pathology to multiple sclerosis disability: a study with magnetisation transfer ratio analysis. J Neurol Neurosurg Psychiatry 69:723–727

85. Iannucci G, Tortorella C, Rovaris M et al (2000) Prognostic value of MR and MTI findings at presentation in patients with clinically isolated syndromes suggestive of MS. Am J NeuroRadiol 21:1034–1038

86. Kaiser JS, Grossman RI, Polansky M et al (2000) Magnetization transfer histogram analysis of monosymptomatic episodes of neurologic dysfunction: Preliminary findings. Am J NeuroRadiol 21:1043–1047

87. Traboulsee A, Dehmeshki J, Brex PA et al (2002) Normal-appearing brain tissue MTR histograms in clinically isolated syndromes suggestive of MS. Neurology 59:126–128

88. Gallo A, Rovaris M, Benedetti B et al (2007) A brain magnetization transfer MRI study with a clinical follow up of about four years in patients with clinically isolated syndromes suggestive of multiple sclerosis. J Neurol 254(1):78–83

89. Fernando KT, Tozer DJ, Miszkiel KA et al (2005) Magnetization transfer histograms in clinically isolated syndromes suggestive of multiple sclerosis. Brain 128:2911–2925

90. van Buchem MA, Grossman RI, Armstrong C, Polansky M et al (1998) Correlation of volumetric magnetization transfer imaging clinical data in MS. Neurology 50:1609–117

91. Iannucci G, Minicucci L, Rodegher M, Sormani MP et al (1999) Correlations between clinical and MRI involvement in multiple sclerosis: assessment using T(1), T(2) and MT histograms. J Neurol Sci 171(2):121–129

92. Cercignani M, Bozzali M, Iannucci G et al (2001) Magnetization transfer ratio and mean diffusivity of normal appearing white and grey matter from patients with multiple sclerosis. J Neurol Neurosurg Psychiatry 70:311–317

93. Ciccarelli O, Werring DJ, Wheeler-Kingshott CA et al (2001) Investigation of MS normal appearing brain using diffusion tensor MRI with clinical correlations. Neurology 56:926–933

94. Caramia F, Pantano P, Di legge S et al (2002) A longitudinal study of MR diffusion changes in normal appearing white matter of patients with early multiple sclerosis. Magn Reson Imaging 20:383–388

95. Gallo A, Rovaris M, Riva R, Ghezzi A et al (2005) Diffusion-tensor magnetic resonance imaging detects normal-appearing white matter damage unrelated to short-term disease activity in patients at the earliest clinical stage of multiple sclerosis. Arch Neurol 62(5):803–808

96. Arnold DL, Matthews PM, Francis G, Antel J (1990) Proton magnetic resonance spectroscopy of human brain in vivo in the evaluation of multiple sclerosis: assessment of the load of disease. Magn Reson Med 14:154–159

97. Fu L, Matthews PM, De Stefano N et al (1998) Imaging axonal damage of normal appearing white matter in multiple sclerosis. Brain 121:103–113

98. De Stefano N, Matthews PM, Fu L et al (1998) Axonal damage correlates with disability in patients with relapsing remitting multiple sclerosis: results of a longitudinal MR spectroscopy study. Brain 121:1469–1477

99. Rocca MA, Cercignani M, Iannucci G, Comi G et al (2000) Weekly diffusion-weighted imaging of normal-appearing white matter in MS. Neurology 55:882–884

100. Brenner RE, Munro PMG, Williams SCR, et al (1993) Abnormal neuronal mitochondria: a cause of reduction in NA in demyelinating disease. In: Proceedings of the SMRM, Amsterdam, p 281

101. Rovaris M, Gambini A, Gallo A et al (2005) Axonal injury in early multiple sclerosis is irreversible and independent of the short-term disease evolution. Neurology 65 (10):1626–1630

102. De Stefano N, Narayanan S, Francis Gs et al (2001) Evidence of axonal damage in the early stages of MS and its relevance to disability. Arch Neurol 58:65–70

103. Fernando KT, McLean MA, Chard DT et al (2004) Elevated white matter myo-inositol in clinically isolated syndromes suggestive of multiple sclerosis. Brain 127:1361–1369
104. Wattjes MP, Harzheim M, Lutterbey GG et al (2008) Prognostic value of high-field proton magnetic resonance spectroscopy in patients presenting with clinically isolated syndromes suggestive of multiple sclerosis. Neuroradiology 50(2):123–129
105. Calabrese M, Filippi M, Rovaris M et al (2009) Evidence for relative cortical sparing in benign multiple sclerosis: a longitudinal magnetic resonance imaging study. Mult Scler 15 (1):36–41
106. Roosendaal SD, Moraal B, Pouwels PJ et al (2009) Accumulation of cortical lesions in MS: relation with cognitive impairment. Mult Scler 15(6):708–714
107. Calabrese M, Agosta F, Rinaldi F et al (2009) Cortical lesions and atrophy associated with cognitive impairment in relapsing-remitting multiple sclerosis. Arch Neurol 66 (9):1144–1150
108. Calabrese M, Rocca MA, Atzori M et al (2009) Cortical lesions in primary progressive multiple sclerosis: a 2-year longitudinal MR study. Neurology 72(15):1330–1336
109. Calabrese M, Rocca MA, Atzori M et al (2010) A 3-year magnetic resonance imaging study of cortical lesions in relapse-onset multiple sclerosis. Ann Neurol 67(3):376–383
110. Calabrese M, Rinaldi F, Seppi D et al (2011) Cortical diffusion-tensor imaging abnormalities in multiple sclerosis: a 3-year longitudinal study. Radiology 261(3):891–898
111. Fisniku LK, Altmann DR, Cercignani M et al (2009) Magnetization transfer ratio abnormalities reflect clinically relevant grey matter damage in multiple sclerosis. Mult Scler 15(6):668–677
112. Sharma J, Zivadinov R, Jaisani Z, Fabiano AJ, Singh B, Horsfield MA, Bakshi R (2006) A magnetization transfer MRI study of deep gray matter involvement in multiple sclerosis. J Neuroimaging 16(4):302–310
113. Amato MP, Portaccio E, Stromillo ML et al (2008) Cognitive assessment and quantitative magnetic resonance metrics can help to identify benign multiple sclerosis. Neurology 71(9):632–638
114. Penny S, Khaleeli Z, Cipolotti L et al (2010) Early imaging predicts later cognitive impairment in primary progressive multiple sclerosis. Neurology 74(7):545–552
115. Khaleeli Z, Altmann DR, Cercignani M et al (2008) Magnetization transfer ratio in gray matter: a potential surrogate marker for progression in early primary progressive multiple sclerosis. Arch Neurol 65(11):1454–1459
116. Penny S, Khaleeli Z, Cipolotti L, Thompson A, Ron M (2010) Early imaging predicts later cognitive impairment in primary progressive multiple sclerosis. Neurology 74(7):545–552
117. Agosta F, Rovaris M, Pagani E et al (2006) Magnetization transfer MRI metrics predict the accumulation of disability 8 years later in patients with multiple sclerosis. Brain 129:2620–2627
118. Bozzali M, Cercignani M, Sormani MP et al (2002) Quantification of brain gray matter damage in different MS phenotypes by use of diffusion tensor MR imaging. AJNR Am J Neuroradiol 23:985–988
119. Rovaris M, Bozzali M, Iannucci G et al (2002) Assessment of normal-appearing white and gray matter in patients with primary progressive multiple sclerosis: a diffusion-tensor magnetic resonance imaging study. Arch Neurol 59:1406–1412
120. Fabiano AJ, Sharma J, Weinstock-Guttman B et al (2003) Thalamic involvement in multiple sclerosis: a diffusion-weighted magnetic resonance imaging study. J Neuroimag 13:307–314
121. Hasan KM, Halphen C, Kamali A et al (2009) Caudate nuclei volume, diffusion tensor metrics, and T(2) relaxation in healthy adults and relapsing-remitting multiple sclerosis patients: implications for understanding gray matter degeneration. J Magn Reson Imaging 29(1):70–77
122. Oreja-Guevara C, Rovaris M, Iannucci G et al (2005) Progressive grey matter damage in patients with relapsing-remitting MS: a longitudinal diffusion tensor MRI study. Arch Neurol 62:578–584

123. Rovaris M, Gallo A, Valsasina P et al (2005) Short-term accrual of gray matter pathology in patients with progressive multiple sclerosis: An in vivo study using diffusion tensor MRI. Neuroimage 24:1139–1146

124. Rovaris M, Judica E, Gallo A et al (2006) Grey matter damage predicts the evolution of primary progressive multiple sclerosis at 5 years. Brain 129(Pt 10):2628–2634

125. Rovaris M, Iannucci G, Falautano M et al (2002) Cognitive dysfunction in patients with mildly disabling relapsing-remitting multiple sclerosis: an exploratory study with diffusion tensor MR imaging. J Neurol Sci 195(2):103–109

126. Benedict RH, Bruce J, Dwyer MG et al (2007) Diffusion-weighted imaging predicts cognitive impairment in multiple sclerosis. Mult Scler 13(6):722–730

127. Kapeller P, McLean MA, Griffin CM et al (2001) Preliminary evidence for neuronal damage in cortical grey matter and normal appearing white matter in short duration relapsing-remitting multiple sclerosis: a quantitative MR spectroscopic imaging study. J Neurol 248:131–138

128. Sarchielli P, Presciutti O, Tarduci R, Gobbi G et al (2002) Localized ^{1}H magnetic resonance spectroscopy in mainly cortical gray matter of patients with multiple sclerosis. J Neurol 249:902–910

129. Chard DT, Griffin CM, McLean MA et al (2002) Brain metabolite changes in cortical grey and normal-appearing white matter in clinically early relapsing-remitting multiple sclerosis. Brain 125:2342–2352

130. Sharma R, Narayana PA, Wolinsky JS (2001) Grey matter abnormalities in multiple sclerosis: proton magnetic resonance spectroscopic imaging. Mult Scler 7:221–226

131. Cifelli A, Arridge M, Jezzard P, Esiri MM, Palace J, Matthews PM (2002) Thalamic neurodegeneration in multiple sclerosis. Ann Neurol 52:650–653

132. Gonen O, Viswanathan AK, Catalaa I, Babb J et al (1998) Total brain N-acetylaspartate concentration in normal, age-grouped females: quantitation with non-echo proton NMR spectroscopy. Magn Reson Med 40:684–689

133. Gonen O, Catalaa I, Babb JS, Ge Y et al (2000) Total brain N-acetylaspartate: a new measure of disease load in MS. Neurology 54:15–19

134. Pulizzi A, Rovaris M, Judica E et al (2007) Determinants of disability in multiple sclerosis at various disease stages: a multiparametric magnetic resonance study. Arch Neurol 64(8):1163–1168

135. Benedetti B, Rovaris M, Rocca MA et al (2009) In-vivo evidence for stable neuroaxonal damage in the brain of patients with benign multiple sclerosis. Mult Scler 15(7):789–794

136. Rovaris M, Gallo A, Falini A et al (2005) Axonal injury and overall tissue loss are not related in primary progressive multiple sclerosis. Arch Neurol 62(6):898–902

137. De Stefano N, Matthews PM, Filippi M et al (2003) Evidence of early cortical atrophy in MS: relevance to white matter changes and disability. Neurology 60(7):1157–1162

138. Sanfilipo MP, Benedict RH, Sharma J et al (2005) The relationship between whole brain volume and disability in multiple sclerosis: a comparison of normalized gray vs. white matter with misclassification correction. Neuroimage 26(4):1068–1077

139. Zivadinov R, Leist TP (2005) Clinical-magnetic resonance imaging correlations in multiple sclerosis. J Neuroimaging 15(4 Supp):10 S–21S, Review

140. Tedeschi G, Lavorgna L, Russo P et al (2005) Brain atrophy and lesion load in a large population of patients with multiple sclerosis. Neurology 65(2):280–285

141. Valsasina P, Benedetti B, Rovaris M et al (2005) Evidence for progressive gray matter loss in patients with relapsing-remitting MS. Neurology 65(7):1126–1128

142. Sastre-Garriga J, Ingle GT, Chard DT et al (2005) Grey and white matter volume changes in early primary progressive multiple sclerosis: a longitudinal study. Brain 128:1454–1460

143. Amato MP, Bartolozzi ML, Zipoli V et al (2004) Neocortical volume decrease in relapsing-remitting MS patients with mild cognitive impairment. Neurology 63(1):89–93

144. Amato MP, Portaccio E, Goretti B et al (2007) Association of neocortical volume changes with cognitive deterioration in relapsing-remitting multiple sclerosis. Arch Neurol 64 (8):1157–1161
145. Sanfilipo MP, Benedict RH, Weinstock-Guttman B et al (2006) Gray and white matter brain atrophy and neuropsychological impairment in multiple sclerosis. Neurology 66(5):685–692
146. Rao SM (1995) Neuropsychology of multiple sclerosis. Curr Opin Neurol 8(3):216–220
147. Calabrese M, Rinaldi F, Mattisi I et al (2010) Widespread cortical thinning characterizes patients with MS with mild cognitive impairment. Neurology 74(4):321–328
148. Bermel RA, Bakshi R, Tjoa C et al (2002) Bicaudate ratio as a magnetic resonance imaging marker of brain atrophy in multiple sclerosis. Arch Neurol 59(2):275–280
149. Houtchens MK, Benedict RH, Killiany R et al (2007) Thalamic atrophy and cognition in multiple sclerosis. Neurology 69(12):1213–1223
150. Sicotte NL, Kern KC, Giesser BS et al (2008) Regional hippocampal atrophy in multiple sclerosis. Brain 131(Pt 4):1134–1141
151. Krupp LB, Serafin DJ, Christodoulou C (2010) Multiple sclerosis-associated fatigue. Expert Rev Neurother 10(9):1437–1447, Review
152. Tedeschi G, Dinacci D, Lavorgna L et al (2007) Correlation between fatigue and brain atrophy and lesion load in multiple sclerosis patients independent of disability. J Neurol Sci 263(1–2):15–19
153. Bakshi R, Czarnecki D, Shaikh ZA et al (2000) Brain MRI lesions and atrophy are related to depression in multiple sclerosis. Neuroreport 11(6):1153–1158
154. Zorzon M, Zivadinov R, Nasuelli D et al (2002) Depressive symptoms and MRI changes in multiple sclerosis. Eur J Neurol 9(5):491–496
155. Feinstein A, Roy P, Lobaugh N et al (2004) Structural brain abnormalities in multiple sclerosis patients with major depression. Neurology 62(4):586–590
156. Ceccarelli A, Rocca MA, Pagani E et al (2008) A voxel-based morphometry study of grey matter loss in MS patients with different clinical phenotypes. Neuroimage 42(1):315–322
157. Morgen K, Sammer G, Courtney SM et al (2006) Evidence for a direct association between cortical atrophy and cognitive impairment in relapsing-remitting MS. Neuroimage 30(3):891–898
158. Sepulcre J, Masdeu JC, Goñi J et al (2009) Fatigue in multiple sclerosis is associated with the disruption of frontal and parietal pathways. Mult Scler 15(3):337–344
159. Andreasen AK, Jakobsen J, Soerensen L et al (2010) Regional brain atrophy in primary fatigued patients with multiple sclerosis. Neuroimage 50(2):608–615
160. Chen JT, Narayanan S, Collins DL et al (2004) Relating neocortical pathology to disability progression in multiple sclerosis using MRI. Neuroimage 23(3):1168–1175
161. Charil A, Dagher A, Lerch JP et al (2007) Focal cortical atrophy in multiple sclerosis: Relation to lesion load and disability. Neuroimage 34(2):509–517
162. Ramasamy DP, Benedict RH, Cox JL et al (2009) Extent of cerebellum, subcortical and cortical atrophy in patients with MS: A case–control study. J Neurol Sci 282(1–2):47–54
163. Calabrese M, Atzori M, Bernardi V et al (2007) Cortical atrophy is relevant in multiple sclerosis at clinical onset. J Neurol 254(9):1212–1220
164. Pellicano C, Gallo A, Li X et al (2010) Relationship of cortical atrophy to fatigue in patients with multiple sclerosis. Arch Neurol 67(4):447–453
165. Filippi M, Rocca MA (2009) Functional MR imaging in multiple sclerosis. Neuroimaging Clin N Am 19(1):59–70, Review
166. Genova HM, Sumowski JF, Chiaravalloti N et al (2009) Cognition in multiple sclerosis: a review of neuropsychological and fMRI research. Front Biosci 14:1730–1744, Review
167. Fox MD, Raichle ME (2007) Spontaneous fluctuations in brain activity observed with functional magnetic resonance imaging. Nat Rev Neurosci 8(9):700–711, Review
168. Damoiseaux JS, Rombouts SA, Barkhof F et al (2006) Consistent resting-state networks across healthy subjects. Proc Natl Acad Sci USA 103(37):13848–13853

169. Raichle ME, MacLeod AM, Snyder AZ et al (2001) A default mode of brain function. Proc Natl Acad Sci USA 98:676–682

170. Greicius MD, Krasnow B, Reiss AL et al (2003) Functional connectivity in the resting brain: A network analysis of the default mode hypothesis. Proc Natl Acad Sci USA 100(1):253–258

171. Vanhaudenhuyse A, Noirhomme Q, Tshibanda LJ et al (2010) Default network connectivity reflects the level of consciousness in non-communicative brain-damaged patients. Brain 133 (Pt1):161–171

172. Rocca MA, Valsasina P, Absinta M et al (2010) Default-mode network dysfunction and cognitive impairment in progressive MS. Neurology 74(16):1252–1259

173. Bonavita S, Gallo A, Sacco R et al (2011) Distributed changes in default-mode resting-state connectivity in multiple sclerosis. Mult Scler 17(4):411–422

174. Roosendaal SD, Schoonheim MM, Hulst HE et al (2010) Resting state networks change in clinically isolated syndrome. Brain 133(Pt 6):1612–1621

175. Schoonheim MM, Geurts JJ, Barkhof F (2010) The limits of functional reorganization in multiple sclerosis. Neurology 74(16):1246–7

176. Prinster A, Quarantelli M, Orefice G et al (2006) Grey matter loss in relapsing-remitting multiple sclerosis: A voxel-based morphometry study. Neuroimage 29(3):859–867

Therapy of Multiple Sclerosis

9

Alessandro d'Ambrosio and Simona Bonavita

Currently there are no therapies able to heal Multiple Sclerosis (MS). However, depending on the stage and course of the disease, there are treatments that can efficiently treat relapses, reduce relapse rate and, partially, disability progression, and improve some symptoms. This means that we are able to change the overall course of the disease.

There are essentially three different kinds of treatment: relapse, disease modifying and symptomatic therapy. Along with pharmacological treatments, physical rehabilitation aims to preserve patients' autonomy, delay the worsening of symptoms and prevent complications.

9.1   Relapse Therapy

The latest European Federation of Neurological Societies guidelines recommend, as first line relapse therapy, intravenous or oral administration of Methylprednisolone (MP) 500–1,000 mg/die for 5 days. There is also a trend towards intravenous administration of MP 1,000 mg for 3 days followed by an oral steroid tapering with Prednisone. Some patients refractory to MP therapy can be treated with plasma exchange [1, 2].

9.2   Disease-Modifying Therapy

This kind of treatment essentially aims to prevent or reduce the number of relapses, to reduce the severity of relapses, and to decrease disability progression. It is based on immunomodulatory and immunosuppressive drugs.

A. d'Ambrosio • S. Bonavita (✉)
Second University of Naples, Naples, Italy
e-mail: simona.bonavita@unina2.it

U. Nocentini et al. (eds.), *Neuropsychiatric Dysfunction in Multiple Sclerosis*,
DOI 10.1007/978-88-470-2676-6_9, © Springer-Verlag Italia 2012

9.2.1 Immunomodulatory Drugs

9.2.1.1 β Interferon (IFNβ)

This immunomodulatory drug has been proved to be effective in the treatment of relapsing-remitting forms of MS; its efficacy in secondary progressive forms is more controversial [3, 4]. An early start to IFNβ therapy reduces relapse rate, disease severity, and new lesion development, and can delay disability progression. Currently three IFNβ formulations have been approved for relapse prevention: intramuscular IFNβ-1a, subcutaneous IFNβ-1a, and subcutaneous IFNβ-1b. These drugs have been successfully used in the last 15 years both in patients with clinically defined MS and in patients with a clinically isolated syndrome (CIS). However, the exact molecular mechanism of the drug's therapeutic effect is not yet clear. The known mechanisms include: regulation of T-lymphocyte activation and of immune cell proliferation, autoreactive T-cell apoptosis, IFNγ antagonism, anti-inflammatory cytokine induction, inhibition of the transit of immune cells through the blood-brain barrier, and possible anti-viral activity.

Very frequent side effects are a flu-like syndrome (fever, chills, joint pain, malaise, headache, myalgia) and local reactions in the injection site.

These interferon formulations can induce the production of antibodies against thyroglobulin and thyroid peroxidase without significant thyroid dysfunction; however, autoimmune thyroiditis may rarely occur, with variable consequences [5–7]. During interferon therapy, liver enzymes may rise to levels that require either dose reduction or therapy discontinuation, which are generally followed by a quick normalization [7]; although it is a rare occurrence, the possibility of a lethal hepatic impairment has been reported [8]. Sometimes there may be a slight reduction of the three series of blood cells (most frequently leukocytes). Monitoring of hematology, thyroid and liver functionality is therefore recommended during interferon therapy.

A very controversial topic is the possible depressant effect of interferon on mood, with episodes of major depression including high suicide risk.

In terms of efficacy, clinical trials have shown a positive effect on the disease, with a relapse rate decrease of 30% and a reduction of disease activity on Magnetic Resonance Imaging (MRI) varying between 50% and 75% [9].

A high probability of developing a new relapse (40–45%) and new MRI lesions (90%) within two years after the onset of symptoms has been demonstrated in CIS patients with clinically silent MRI lesions. Studies on the three types of IFNβ in CIS patients have shown a decrease in the risk of developing new clinically silent MRI lesions, a time prolongation between onset and second clinical relapse, and decreased disability progression [10–13].

In patients with secondary progressive forms (SP) of MS, different IFNβ formulations may have slight efficacy only in patients with clinical relapses [14, 15].

Five to 25% of patients treated with IFNβ may develop neutralizing antibodies (NABs) against IFNβ that significantly reduce treatment efficacy. Generally speaking, NABs appear within 6–18 months from treatment start; accordingly, a plasmatic NAB dosage is recommended after 12 and 24 months of therapy and when there is no response. In patients with high titre NABs persisting over several dosages, the therapy must be discontinued to switch to other disease-modifying drugs [16].

9.2.1.2 Glatiramer Acetate

Glatiramer acetate (GA, Copaxone®, also known as Copolymer-1) is the first non-interferon drug therapy approved by the Food and Drug Administration for the treatment of relapsing-remitting forms of MS. Over the years, many mechanisms of action have been attributed to GA, both in MS and in its animal model, Experimental Autoimmune Encephalomyelitis. The implicated mechanisms justifying the immunomodulatory effect of GA are: suppression of cell proliferation, induction of tolerance, expansion of T-cell regulatory population and alterations of antigen-presenting cells. A pivotal trial on relapsing-remitting forms of MS showed a 30% reduction of relapse rate and a significant reduction of disease activity on MRI [17]. Also in patients with a CIS, GA proved to be effective in delaying conversion to clinically defined MS [18]. Comparative studies between GA, IFNβ-1a, and IFNβ-1b showed a comparable efficacy in the various parameters evaluated [19, 20].

In all the clinical trials, the most frequent adverse reactions reported by the majority of patients are in the injection site. Moreover, a post-injection reaction was described, characterized by one or more of the following symptoms: vasodilatation, chest pain, dyspnea, palpitations or tachycardia [21].

9.2.1.3 Monoclonal Antibodies

This is a group of highly selective drugs whose specific target is cell-surface molecules. Currently many monoclonal antibodies (Rituximab. Alemtuzumab, Daclizumab, Ocrelizumab) are under investigation in MS treatment. However, at present only Natalizumab has been approved for marketing. This monoclonal antibody is a selective inhibitor of an adhesion molecule and binds to the α4 subunit of human integrins, highly expressed on the surface of leukocytes, except for neutrophils. Its therapeutic effect is mainly due to its inhibition of activated leukocytes adhering to the endothelium, thus preventing inflammatory cell migration to the central nervous system. This monoclonal antibody is indicated as monotherapy in patients with a relapsing-remitting form of MS; more specifically, in Europe it has been approved for the treatment of patients in whom a course of IFNβ therapy has failed or for patients affected by a relapsing-remitting form of MS with high relapse rate or rapid disability progression. The AFFIRM pivotal trial on Natalizumab monotherapy [22] showed a 70% reduction of relapse rate, a 42–54% reduction of disability progression and a significant reduction (90%) of disease activity on MRI (determined as a reduction of gadolinium-enhancing lesions). In Natalizumab-treated patients a variable risk of opportunistic infections is reported, mostly progressive multifocal leukoencephalopathy (PML). The evaluation of past immunosuppressive treatments, number of Natalizumab infusions, and detection of anti-JCV antibodies may provide a means to stratify MS patients considering or receiving this treatment and assess their risk of developing PML. In particular, in JCV-positive patients who have had prior treatment with immunosuppressive drugs and two years of Natalizumab infusions the risk of PML is significantly increased [23].

Natalizumab too may induce an immune reaction with a persistent production of NABs; after 6 and 12 months of therapy and where there are infusion-related adverse events or no response, a plasmatic NAB dosage is recommended [24].

9.2.2 Immunosuppressive Drugs

In MS patients there is much evidence to suggest that an early and aggressive intervention may improve the short and long term evolution of the disease. Cytotoxic drugs may offer many advantages, especially when used as induction therapy or add-on therapy to immunomodulatory drugs. The common action mechanism for all cytotoxic drugs is immunosuppression. Currently the only cytotoxic drug approved for MS treatment is Mitoxantrone. However, other drugs have been used to a limited extent or are in the testing phase.

9.2.2.1 Mitoxantrone

This is an anti-neoplastic drug able to bind DNA, inducing strand breakage, and inhibiting DNA and RNA biosynthesis. Its therapeutic effect is due to activity inhibition of B-T lymphocytes and macrophage proliferation. It is indicated in patients affected by relapsing-remitting forms with high relapse rate or rapid disability progression, and in patients affected by SP forms with signs of inflammatory activity (clinical relapses, response to steroid therapy, gadolinium-enhancing lesions) [25]. Controlled clinical trials showed a 60–70% reduction of relapse rate, a decrease of disability progression and disease activity on MRI [26]. The Mitoxantrone treatment can lead to serious side effects, especially cardiotoxicity, myelosuppression, and rarely leukemia. Due to the potential cardiotoxic effect the total maximum dose is restricted to 120–140 mg/m^2 of skin surface, and echocardiographic exams before and during treatment are recommended.

9.2.2.2 Azathioprine

Until the mid 90s, before the introduction of interferons, this drug was frequently used for MS treatment. Recently, its efficacy and tolerability have been revaluated on Cochrane based methodology due to its low cost and potential efficacy in reducing relapse rate and disability progression [27]. The meta-analysis indicates that azathioprine reduces relapses during the first, second, and third years of follow-up, and reduces disease progression during the first 2–3 years.

9.2.2.3 Cyclophosphamide

This is an antineoplastic alkylating and cytostatic drug belonging to the group of nitrogen mustards. It expresses a selective effect in immune response, such as activity suppression of T CD4+ Th1 lymphocytes (pro-inflammatory cells) and response increase of T CD4+ Th2 lymphocytes (anti-inflammatory cells); both mechanisms are involved in the Cyclophosphamide therapeutic effect. However, widespread use is hampered by the drug's toxic effect, especially at bladder level, and the risk of cancer.

Generally speaking it is indicated in relapsing-remitting forms, characterized by close relapses with rapid disability progression, or in the first period of secondary progressive forms, in patients non-responsive to other drugs.

9.2.2.4 Methotrexate

This a very powerful antineoplastic antimetabolite drug, a structural antagonist to folic acid and an inhibitor of human dihydrofolate reductase, an enzyme involved in the synthesis of essential molecules like DNA and RNA. In the treatment of MS, Methotrexate may theoretically have a beneficial effect on relapse rate and disability progression. However, before we can draw conclusions, further studies are needed on relapsing-remitting and progressive forms.

9.2.2.5 New Oral Therapies

The introduction of new oral drugs aims not only to remove injective therapy, poorly tolerated in the long-term, but also to provide more effective treatment, in terms of relapse rate, disability progression and lesion load reduction.

Five oral drugs are currently under evaluation in phase II/III clinical trials.

Fingolimod, an atypical immunosuppressive drug, reduces relapse rate by 54% and the risk of disability progression by 37% compared to placebo. Treatment with Fingolimod has also resulted in statistically significant reductions in brain lesion activity and reduced loss of brain volume as measured by MRI. Fingolimod-related adverse events include a transient, generally asymptomatic heart rate reduction, infrequent transient atrio-ventricular conduction block, mild (1–3 mmHg) blood pressure increase, macular edema, and asymptomatic, reversible elevation of liver enzymes [28, 29]. This drug is already on the market in the United States and in some European countries.

To date seven unexplained deaths have been reported, four of them sudden. In addition, other reports include three deaths due to heart attack and one due to disruption of the heart rhythm. Currently it is not clear whether these were directly caused by Fingolimod.

Fingolimod is currently under review by the European Medicines Agency following cases of death and serious cardiovascular events in patients who had recently started treatment with the drug. Meanwhile the Agency's Committee for Medicinal Products for Human Use is advising healthcare professionals to intensify monitoring of patients after the first dose administration.

Laquinimod, an immunomodulatory drug in phase II trial, reduces the number of active lesions at the highest tested dose (0.6 mg/dl). Cladribine, another immunosuppressive drug, induces a >50% reduction of relapse rate and a >70% reduction of gadolinium-enhancing lesions at both tested dosages; however, this drug probably will not be marketed due to methodological errors in the trial procedures.

Oral Fumarate, with immunomodulatory and antioxidant properties, reduces the number of lesions and is in phase III trial. Finally, Teriflunomide, a phase II trial immunomodulatory drug, significantly reduces MRI lesion activity and relapse rate.

9.2.2.6 Chronic Cerebrospinal Venous Insufficiency (CCSVI)

CCSVI as a potential etiopathogenic entity in MS has recently been suggested and has gained significant attention. Over the past few years, Zamboni and colleagues were the first to propose and explore this novel and controversial idea. The theory of CCSVI suggests that MS may be causally related to an inflammatory or immune reaction to iron that accumulates in the central nervous system secondary to insufficiency of cervical and cerebral venous blood vessels. On the basis of the available literature, it is currently unclear whether CCSVI exists as a pathologic entity or as an anatomic variant in patients with MS. It is even less substantiated that there is a causal association between CCSVI and MS. Pending further investigations, no therapeutic interventions involving treatment of CCSVI should be recommended for patients with MS outside of controlled clinical trials [30].

9.3   Symptomatic Therapy

Specific treatment of symptoms is an essential part of MS patient management. The aim of symptomatic treatment is to remove or reduce symptoms that worsen patients' quality of life.

9.3.1   Spasticity

Treatment aims are: improvement of motor function, reduction of pain, avoidance of complications, and removal of factors (urinary infections, constipation or fever) that could trigger or worsen spasticity.

9.3.1.1 Physiotherapy

This is generally accepted as the basic treatment for spasticity, although only a few controlled trials have been performed in MS patients.

9.3.1.2 Pharmacological Treatment

The most common oral antispastic drugs are Baclofen and Tizanidine. Many trials have shown the efficacy of continuous intrathecal Baclofen infusion for MS patients with serious spinal or supraspinal spasticity, significantly reducing muscle tone and spasm frequency, with a potential improvement of life quality. Unfortunately, adverse events like muscle fatigue, headache, consciousness disorders, infections or catheter dislocation may represent a significant obstacle when using this therapy.

Botulinum toxin is an important new weapon, especially for focal limb spasticity.

9.3.2 Fatigue

More than 75% of patients suffer from extreme fatigue that usually increases during the day. MS fatigue may limit professional activity and social life, and is sometimes one of the major causes of disability. The aims of the treatment are fatigue reduction and facilitation of normal social and occupational activities.

Transient cooling of body or limbs using cold packs, cold baths or air conditioning may improve gait, postural stability, fatigue and lower-limb muscle strength.

Sulphate Amantadine leads to a moderate improvement of subjective fatigue, ability to concentrate and memory [31].

Fampridine (Dalfampridine, 4-Aminopyridine, 4-AP) and 3,4-Diaminopyridine (DAP): Fampridine seems to be more effective than 3,4-DAP in improving temperature-related symptoms. Its use is restricted by a narrow therapeutic window; adverse events include nausea and rarely, seizures [32].

The efficacy of fampridine in improving walking was established in two double-blind studies in which patients with multiple sclerosis were randomised to receive placebo or 10 mg prolonged-release fampridine, twice daily. Completion of a timed 25-ft walk was used as the primary efficacy endpoint in both studies. In one study (n = 296), 34.8% of patients taking fampridine responded to the treatment, with an average increase in walking speed of 26.3% vs 5.3% on placebo (p < 0.001). The other study (n = 237) showed similar results with 42.9% responding to treatment and an increase in walking speed of 25.3% vs 7.8% (p < 0.001) [32, 33].

9.3.3 MS Related Pain

The frequency of clinically relevant pain is reported in 29–86% of patients. The aim of the therapy is pain reduction for an improved quality of life. An accurate definition of pain is crucial for choosing the correct treatment.

Chronic pain, directly related to the disease, often occurs asymmetrically or bilaterally in the form of an uncomfortable burning sensation and limb or trunk dysesthesia. This neuropathic pain, which may occur in many neurological diseases, is relieved by tricyclic antidepressants such as amitriptyline, by SNRI antidepressants such as duloxetine, or antiepileptic drugs such as carbamazepine. Good results can also be obtained with Pregabalin, Lamotrigine and Gabapentin.

In some cases, pain can be an indirect consequence of the disease, for example pain due to joint and muscle overload; patients should be encouraged to work actively to achieve a better posture.

Local pain during IFNβ or Glatiramer acetate treatment can be prevented by the application of cold packs before and after injection. Flu-like symptoms with muscular pain can be relieved by Acetaminophen, Ibuprofen, or other NSAIDs.

9.3.4   Bladder Symptoms

Bladder disorders occur in more than 80% of patients during the disease course and often significantly worsen the quality of life. The aims of the treatment are bladder function improvement and prevention of complications.

Patients should be encouraged to keep a urination diary, and to drink properly and regularly during the day. Biofeedback and pelvic exercises are suggested to reduce urinary urgency and incontinence.

Many trials showed the positive effect of anticholinergic drugs, such as Oxybutynin or Tolterodine, in reducing urgency and incontinence. Side effects of anticholinergic drugs can be attenuated by symptomatic drugs or slow-release preparations. Alpha-blocker agents such as Alfuzosine and Tamsulosine reduce non-obstructive urinary retention. Oral Baclofen has positive effects on spastic or dyssynergic sphincters. Urinary infections are very common in patients with bladder disorders and should be treated for at least 10 days with a specific antibiotic therapy. Sterile catheterization 4–6 times a day is the best treatment for retention due to hyperreflexic bladder and sphincter hypertonia; however, bacterial infections are a very common risk.

9.3.5   Sexual Dysfunctions

During the course of MS sexual dysfunctions occur in more than 80% of patients, often related to bladder disorders. Men are affected more frequently than women. The aim of treatment is to regulate the patient's sexual activity. The most commonly used drugs are 5-phosphodyesterase inhibitors such as Sildenafil, Tadalafil, Vardenafil. For female sexual disorders (anorgasmia, vaginal dryness, dyspareunia, decreased libido) a primary role can be played by appropriate psychological or neurological counseling. For men too, psychological support can be part of an interdisciplinary approach.

9.3.6   Ataxia and Tremor

The aim of the treatment is to reduce ataxia, especially when it interferes with daily activities. Cornerstones of the treatment are physiotherapy and occupational therapy. Most of the published literature on medical treatment consists of case reports and uncontrolled open label studies to evaluate the efficacy of some drugs such as Carbamazepine, Propranolol or Ondansetron. These pharmacological treatments can only reduce the tremor component, and are not usually very effective.

9.3.7   Paroxysmal Symptoms

About 10–20% of patients suffer from paroxysmal symptoms such as trigeminal neuralgia or other paroxysmal pains. The main treatment consists of antiepileptic drugs, primarily Carbamazepine, Gabapentin and Pregabalin.

9.3.8 Dysphagia

The frequency ranges from 24% to 55% of patients. Urgent indications of the necessity for active treatment of dysphagia are dehydration, malnutrition and bronchial aspiration. The treatment consists of functional and pharmacological therapy, and other palliative measures such as Percutaneous Endoscopic Gastrostomy. Anticholinergic drugs can reduce hypersalivation. Botulinum toxin, injected in the upper esophageal sphincter, can reduce dysphagia due to increased sphincter tone.

9.3.9 Cognitive Dysfunctions

The only treatment of cognitive disorders we will mention is the use of cholinesterase inhibitors in MS patients; for cognitive rehabilitation, therapy of mood disorders, mania and psychosis see the specific sections.

Trials on the efficacy of cholinesterase inhibitors used to improve cognitive deficits in MS patients are inconclusive, especially as regards long-term therapy.

References

1. Sellebjerg F, Barnes D, Filippini G et al (2005) EFNS guideline on treatment of multiple sclerosis relapses: report of an EFNS task force on treatment of multiple sclerosis relapses. Eur J Neurol 12(12):939–946
2. Weinshenker BG, O'Brien PC, Petterson TM et al (1999) A randomized trial of plasma exchange in acute central nervous system inflammatory demyelinating disease. Ann Neurol 46(6):878–886
3. The IFNB Multiple Sclerosis Study Group and The University of British Columbia MS/MRI Analysis Group (1995) Interferon beta-1b in the treatment of multiple sclerosis: final outcome of the randomized controlled trial. Neurology 45(7):1277–1285
4. PRISMS (Prevention of Relapses and Disability by Interferon beta-1a Subcutaneously in Multiple Sclerosis) Study Group (1998) Randomised double-blind placebo-controlled study of interferon beta-1a in relapsing/remitting multiple sclerosis. Lancet 352(9139): 1498–1504
5. Schwid SR, Goodman AD, Mattson DH (1997) Autoimmune hyperthyroidism in patients with multiple sclerosis treated with interferon beta-1b. Arch Neurol 54(9):1169–1190
6. Rotondi M, Oliviero A, Profice P et al (1998) Occurrence of thyroid autoimmunity and dysfunction throughout a nine-month follow-up in patients undergoing interferon-beta therapy for multiple sclerosis. J Endocrinol Invest 21(11):748–752
7. Durelli L, Ferrero B, Oggero A et al (1999) Autoimmune events during interferon beta-1b treatment for multiple sclerosis. J Neurol Sci 162(1):74–83
8. Yoshida EM, Rasmussen SL, Steinbrecher UP et al (2001) Fulminant liver failure during interferon beta treatment of multiple sclerosis. Neurology 56(10):1416
9. Li DK, Paty DW (1999) Magnetic resonance imaging results of the PRISMS trial: a randomized, double-blind, placebo-controlled study of interferon-beta1a in relapsing-remitting multiple sclerosis. Prevention of relapses and disability by interferon-beta1a subcutaneously in multiple sclerosis. Ann Neurol 46(2):197–206

10. Jacobs LD, Beck RW, Simon JH et al (2000) Intramuscular interferon beta-1a therapy initiated during a first demyelinating event in multiple sclerosis. CHAMPS study group. N Engl J Med 343(13):898–904

11. Comi G, Filippi M, Barkhof F et al (2001) Effect of early interferon treatment on conversion to definite multiple sclerosis: a randomised study. Lancet 357(9268):1576–1582

12. Kappos L, Polman CH, Freedman MS et al (2006) Treatment with interferon beta-1b delays conversion to clinically definite and McDonald MS in patients with clinically isolated syndromes. Neurology 67(7):1242–1249

13. Kappos L, Freedman MS, Polman CH et al (2007) Effect of early versus delayed interferon beta-1b treatment on disability after a first clinical event suggestive of multiple sclerosis: a 3-year follow-up analysis of the BENEFIT study. Lancet 370(9585):389–397

14. Panitch H, Miller A, Paty D et al (2004) Interferon beta-1b in secondary progressive MS: results from a 3-year controlled study. Neurology 63(10):1788–1795

15. Secondary Progressive Efficacy Clinical Trial of Recombinant Interferon-Beta-1a in MS (SPECTRIMS) Study Group (2001) Randomized controlled trial of interferon- beta-1a in secondary progressive MS: Clinical results. Neurology 56(11):1496–1504

16. Sørensen PS, Deisenhammer F, Duda P et al (2005) Guidelines on use of anti-IFN-beta antibody measurements in multiple sclerosis: report of an EFNS Task Force on IFN-beta antibodies in multiple sclerosis. Eur J Neurol 12(11):817–827

17. Johnson KP, Brooks BR, Cohen JA et al (2001) Copolymer 1 reduces relapse rate and improves disability in relapsing-remitting multiple sclerosis: results of a phase III multicenter, double-blind, placebo-controlled trial. Neurology 57(12 Suppl 5):S16–S24

18. Comi G, Martinelli V, Rodegher M et al (2009) Effect of glatiramer acetate on conversion to clinically definite multiple sclerosis in patients with clinically isolated syndrome (PreCISe study): a randomised, double-blind, placebo-controlled trial. Lancet 374(9700):1503–1511

19. O'Connor P, Filippi M, Arnason B et al (2009) 250 microg or 500 microg interferon beta-1b versus 20 mg glatiramer acetate in relapsing-remitting multiple sclerosis: a prospective, randomised, multicentre study. Lancet Neurol 8(10):889–897

20. Mikol DD, Barkhof F, Chang P et al (2008) Comparison of subcutaneous interferon beta-1a with glatiramer acetate in patients with relapsing multiple sclerosis (the REbif vs Glatiramer Acetate in Relapsing MS Disease [REGARD] study): a multicentre, randomised, parallel, open-label trial. Lancet Neurol 7(10):903–914

21. Johnson KP, Brooks BR, Cohen JA et al (2001) Extended use of glatiramer acetate (Copaxone) is well tolerated and maintains its clinical effect on multiple sclerosis relapse rate and degree of disability. Neurology 57((12 Suppl 5)):S46–S53

22. Polman CH, O'Connor PW, Havrdova E et al (2006) A randomized, placebo-controlled trial of natalizumab for relapsing multiple sclerosis. N Engl J Med 354(9):899–910

23. Yousry TA, Major EO, Ryschkewitsch C et al (2006) Evaluation of patients treated with natalizumab for progressive multifocal leukoencephalopathy. N Engl J Med 354(9):924–933

24. Kappos L, Bates D, Hartung HP et al (2007) Natalizumab treatment for multiple sclerosis: recommendations for patient selection and monitoring. Lancet Neurol 6(5):431–441

25. Neuhaus O, Kieseier BC, Hartung HP (2006) Therapeutic role of mitoxantrone in multiple sclerosis. Pharmacol Ther 109(1–2):198–209

26. Millefiorini E, Gasperini C, Pozzilli C et al (1997) Randomized placebo-controlled trial of mitoxantrone in relapsing-remitting multiple sclerosis: 24-month clinical and MRI outcome. J Neurol 244(3):153–159

27. Casetta I, Iuliano G, Filippini G (2009) Azathioprine for multiple sclerosis. J Neurol Neurosurg Psychiatry 80(2):131–132

28. Cohen JA, Barkhof F, Comi G et al (2010) Oral fingolimod or intramuscular interferon for relapsing multiple sclerosis. N Engl J Med 362(5):402–415

29. Kappos L, Radue EW, O'Connor P et al (2010) A placebo-controlled trial of oral fingolimod in relapsing multiple sclerosis. N Engl J Med 362(5):387–401

30. Zivadinov R, Ramanathan M, Dolic K et al (2011) Chronic cerebrospinal venous insufficiency in multiple sclerosis: diagnostic, pathogenetic, clinical and treatment perspectives. Expert Rev Neurother 11(9):1277–1294
31. Peuckmann V, Elsner F, Krumm N et al (2010) Pharmacological treatments for fatigue associated with palliative care. Cochrane Database Syst Rev 11:CD006788
32. Goodman AD, Brown TR, Edwards KR et al (2010) A phase 3 trial of extended release oral dalfampridine in multiple sclerosis. Ann Neurol 68(4):494–502
33. Goodman AD, Brown TR et al (2009) Sustained-release oral fampridine in multiple sclerosis: a randomised, double-blind, controlled trial. Lancet 373(9665):732–738

Rehabilitation in Multiple Sclerosis 10

Ugo Nocentini and Carlo Caltagirone

10.1 General Aspects

The topic of this volume demands a brief discussion of the rehabilitation of patients with MS.

All possible treatment strategies, including rehabilitative ones, must be integrated to improve the wellbeing of these patients. Accordingly, we will give some general information and discuss the topic of rehabilitation in patients with MS. Finally, we will briefly mention cognitive rehabilitation.

As MS is an extremely composite and variable pathology, it requires many different rehabilitative approaches. We must consider the personal characteristics of patients with MS, the instability of their functional situation, the possible involvement of all aspects of neurological functioning and, consequently, of all bodily structures. Therapeutic and rehabilitative approaches must be well timed, individualized, flexible and multi-disciplinary.

In MS, rehabilitation, like other therapies, is very complex. We know that the nervous system has a high degree of structural and functional plasticity which can form a basis for recovery.

If we wish to include rehabilitation in the therapy available for patients with MS, we must be able to evaluate its effectiveness. This involves applying the principles and procedures used to evaluate the effectiveness and safety of any therapy.

Unfortunately, there are no data on the sensitivity to rehabilitation-induced changes of the instruments commonly used to evaluate patients with MS. Lack of a single instrument able to evaluate the many aspects of this illness means that a group of instruments must be used, each of which measures a specific aspect.

There is another important methodological aspect to be considered in evaluating the effectiveness of rehabilitation. For a protocol to be considered valid for

U. Nocentini (✉) • C. Caltagirone
Dipartimento di Neuroscienze, Università degli Studi di Roma "Tor Vergata", Rome, Italy
e-mail: u.nocentini@hsantalucia.it; c.caltagirone@hsantalucia.it

U. Nocentini et al. (eds.), *Neuropsychiatric Dysfunction in Multiple Sclerosis*,
DOI 10.1007/978-88-470-2676-6_10, © Springer-Verlag Italia 2012

establishing therapeutic effectiveness and degree of risk of a specific treatment the following are required: comparison between a group in active therapy and a group subjected to an intervention without therapeutic effects (placebo); random assignment to one group or the other; a double-blind situation in which neither the patient nor the clinician is aware of the nature of the treatment.

Although some researchers believe this is practically impossible in the case of rehabilitation, in reality some studies have shown that it is difficult but possible. Indeed, in pursuing this goal there has been a notable increase in the methodological rigor of studies evaluating the effectiveness of rehabilitative treatments in patients with MS. Much research has been carried out to evaluate the effectiveness of both individual aspects of the rehabilitative process and an entire group of rehabilitative interventions. There is evidence supporting the validity of some components of the rehabilitative process applied individually and of several rehabilitative aspects applied simultaneously. [For a review of general issues in rehabilitation of MS see 1–3].

The economic aspects connected with logistical-organizational aspects have also been evaluated [4]. For several types of rehabilitation (occupational therapy, therapeutic exercises), evidence-based reviews provide an overall assessment of effectiveness [5, 6].

Although not all the questions on the rehabilitation of patients with MS have been completely resolved, some possible future solutions can be predicted: increasing knowledge of the mechanisms of recovery and compensation will lead to more adequate restructuring of rehabilitative programs; effectiveness will be shown by more rigorous methodologies for verifying results; criteria independent of the evaluator, such as those derived from functional neuro-imaging techniques, will be introduced.

To use resources in the best way possible, the characteristics of patients with MS who will benefit most from rehabilitation must be identified. As our knowledge of the illness increases, however, we are finding that motor and cognitive deficits are already present in the precocious phase. Rehabilitation will accordingly also be aimed at this area. For slight deficits, it might be enough to give patients suggestions about how to manage the activities of daily living or how they should carry out physical activity alone to keep their body in better physical condition. For moderate problems, brief rehabilitation programs focusing on compromised areas and integrated with the approach suggested for lighter deficits might be sufficient. In patients with larger deficits, more intensive and articulated programs will be needed. And in cases of severe disability, the highest possible levels of comfort must be guaranteed [7].

Given the many problems that MS patients can encounter, all areas of rehabilitation are potentially useful, and often several different rehabilitative approaches must be adopted simultaneously.

One focus of attention is the development of organizational modalities for providing rehabilitation. Despite the diversity among MS patients, they are able to live in the community for most of the course of their illness, and it is essential that this situation be maintained. Day hospital or outpatient rehabilitation will therefore be the best solutions for most of these patients. Ordinary hospitalization and home care should be reserved for patients who have great physical difficulty leaving the home. The advantages and disadvantages of each option should be discussed with the patient, family members and caregivers to identify the one that offers the best balance. And in the area of health care politics, it will be necessary to see that the various options are feasible.

We need to think in terms of integrating all therapeutic, assistance and social aspects relating to patients with MS. The efforts of all those working to improve the condition of these patients must be directed at integrating care, via multidisciplinary routes and involving all the patient's carers. These focus on the patient and relate to a specific pathological condition and a particular treatment situation. Furthermore, they provide detailed documentation of the process involved in providing therapy. Deviations from the accepted process are documented and analyzed, and analysis of the variations permits the team to review clinical practice, redefine its course and develop more efficient, appropriate and timely therapies [8, 9].

Integration cannot be limited to what is available at the hospital or local level. It must provide a real connection between all factors in play and also identify professionals to organize, support and verify the outcome of this "superstructural" integration. If this route is followed, the patient will necessarily be at the centre of the therapeutic process and those closest to them will become knowledgeable collaborators.

Only a multi-disciplinary team can develop and manage the integrated pathway and verify its effectiveness [10, 11]. Our goal should be to prepare interdisciplinary teams which would ensure more integrated therapies with a common approach for planning, providing therapy and evaluating outcomes.

Finally, rehabilitation should not be separate from other therapies. If interventions are combined, better results will be obtained.

10.2 Cognitive Rehabilitation in Multiple Sclerosis

There are only about 20 studies in the literature of strategies and cognitive rehabilitation programs to treat patients with MS. There are also three reviews. The first is the Cochrane Collaboration Review by Thomas et al. [12], which considered the only five studies presenting the characteristics of randomized and controlled trials before 2007. Nevertheless, it was impossible to carry out a meta analysis of the data of these five studies. On the basis of available data, the authors concluded that it is impossible to judge the effectiveness of cognitive rehabilitation, regardless of the treatment modality used.

The second review [10] considered 16 studies. The authors concluded that if we use an evidence-based principle, we must accept that we are still at the beginning. Indeed, what has been done up to now provides only some basics for future research. Most previous studies concerned learning and memory ability. According to O'Brien et al. [10], in the practical-applicative field it is possible to recommend a story memory technique used in a study by Chiaravalloti et al. [11]. A series of recommendations are proposed in the review: the need for greater methodological rigor, improved modalities of assessment, larger sample sizes, a more precise description of the general characteristics of the population and of the methodology of the intervention, application of methodologies proven effective in treating other pathological conditions and evaluation of the transfer of benefits to activities of daily living.

A third, very recent review was published in the framework of the Cochrane Collaboration by Rosti-Otajärvi and Hämäläinen [13]: this review is specifically devoted to neuropsychological rehabilitation and, following the rules of evidence-based medicine, it has included only randomized controlled and quasi-randomized trials. The authors conclude that there is low level evidence of the positive effects of neuropsychological rehabilitation in MS: specifically, the trials considered show the possibility of improving aspects of memory (memory span, immediate visual memory) and working memory. The limits of existing trials as well as recommendations for future work are also included in the review.

In conclusion, the characteristics of the cognitive disorders caused by MS make it difficult to plan and administer rehabilitation programs and trials whose effectiveness can be demonstrated.

Until now therapists and psychologists have used both computerized programs and direct training. Besides re-educating attention and mnesic functions, which account for the largest part of the intervention, they also treat patients' behavioural disturbances or deficits.

Precise details of rehabilitative methodology are lacking in most studies, with the exception of those that used programs included in the Rehacom system [14–16].

In any case, we agree with the authors of the cited reviews that the available results do not permit definitive conclusions to be drawn. We also agree with their recommendations.

References

1. Clanet MG, Brassat D (2000) The management of multiple sclerosis. Curr Opin Neurol 13:263–270
2. Freeman JA, Thompson AJ (2001) Building an evidence base for multiple sclerosis management: support for physiotherapy. J Neurol Neurosurg Psychiatry 70:147–148
3. Thompson AJ (2001) Symptomatic management and rehabilitation in multiple sclerosis. J Neurol Neurosurg Psychiatry 71(suppl II):ii22–ii27
4. Wiles CM, Newcombe RG, Fuller KJ, Shaw S, Furnival-Doran J, Pickersgill TP (2001) Controlled randomised crossover trial of the effects of physiotherapy on mobility in chronic multiple sclerosis. J Neurol Neurosurg Psychiatry 70:174–179
5. Steultjens EMJ, Dekker J, Bouter LM, Cardol M, Van de Nes JCM, Van den Ende CHM (2003) Occupational therapy for multiple sclerosis. The Cochrane Database of Systematic Reviews. Issue 3: CD003608
6. Rietberg MB, Brooks D, Uitdehaag BMJ, Kwakkel G (2005) Exercise therapy for multiple sclerosis. The Cochrane Database of Systematic Reviews Issue 1: CD003980
7. Rousseaux M, Perennou D (2004) Comfort care in severely disabled multiple sclerosis patients. J Neurol Sci 222:39–48

8. Rossiter D, Thompson AJ (1995) Introduction of integrated care pathways for patients with multiple sclerosis in an inpatient neurorehabilitation setting. Disabil Rehabil 17:443–448

9. Rossiter DA, Edmondson A, Al-Shahi R, Thompson AJ (1998) Integrated care pathways in multiple sclerosis rehabilitation: completing the audit cycle. Mult Scler 4:85–89

10. O'Brien AR, Chiaravalloti N, Goverover Y, Deluca J (2008) Evidenced-based cognitive rehabilitation for persons with multiple sclerosis: a review of the literature. Arch Phys Med Rehabil 89:761–769

11. Chiaravalloti ND, Deluca J, Mooore NB, Ricker JH (2005) Treating learning impairments improves memory performance in multiple sclerosis: a randomized clinical trial. Mult Scler 11:58–68

12. Thomas PW, Thomas S, Hillier C, Galvin K, Baker R (2006). Psychological interventions for multiple sclerosisl. Cochrane Database Syst Rev 25: CD004431

13. Rosti-Otajärvi EM, Hämäläinen PI (2011) Neuropsychological rehabilitation for multiple sclerosis. Cochrane Database of Systematic Reviews Issue 11 Art No: CD009131

14. Mendozzi L, Pugnetti L, Motta A, Barbieri E, Gambini A, Cazzullo CL (1998) Computer assisted memory retraining of patients with multiple sclerosis. Ital J Neurol Sci 19:S431–S432

15. Solari A, Motta A, Mendozzi L et al (2004) Computer-aided retraining of memory and attention in people with multiple sclerosis: a randomized, double-blind controlled trial. J Neurol Sci 222:99–104

16. Tesar N, Bandion K, Baumhackl U (2005) Efficacy of a neuropsychological training programme for patients with multiple sclerosis-a randomised controlled trial. Wien Klin Wochenschr 117:747–754

Part II

Psychiatric Disturbances
in Multiple Sclerosis

Depression and Anxiety 11

Alberto Siracusano, Cinzia Niolu, Lucia Sacchetti, and
Michele Ribolsi

11.1 Epidemiology

Major Depressive Disorder (MDD) is very common in patients with Multiple Sclerosis (MS), with a lifetime prevalence of up to 50 % [1]. In the last 15–20 years various studies have revealed that the annual prevalence of MDD in MS is higher compared both to the general population and people with different chronic medical conditions. In MS patients aged between 18 and 45, an annual prevalence of 25.7 % has been reported [2]. Interestingly, psychological distress in MS patients is associated with neurological disability, but is also present in patients with minimal or no neurological disability. Major Depression and affective disturbances in general are identified as an independent predictor for MS-related quality of life [3].

The fact that suicidal intent is relatively common in MS patients, and that depression in MS patients is often not recognized and is not treated, is of major importance [1, 4]. A 1990 review paper on affective disorders in this kind of patient [5] reported a higher incidence and prevalence of depressive symptoms in MS than in control subjects with different neurological disorders. Minden et al. [6] selected almost randomly 50 MS patients, reporting that 54 % of the sample satisfied the research criteria for Major Depression. Joffe et al. [7] analyzed 100 consecutive patients at an MS clinic in Canada and reported a lifetime prevalence of 47 %. In another study of 221 consecutive patients attending an MS clinic in Vancouver, Sadovnick et al. (1996) [8] reported a lifetime prevalence rate of 50 % using a structured psychiatric interview. Chwastiak et al. (2002) [9] mailed a survey to 1,374 members of the Multiple Sclerosis Association of King County, with a 54 % response rate; they found that 42 % of the study sample had clinically significant

A. Siracusano • C. Niolu • L. Sacchetti • M. Ribolsi (✉)
Psychiatric Division, University Hospital Tor Vergata, Rome, Italy
e-mail: siracusano@med.uniroma2.it; niolu@med.uniroma2.it; luciasa@hotmail.it; mic19812000@yahoo.it

U. Nocentini et al. (eds.), *Neuropsychiatric Dysfunction in Multiple Sclerosis*,
DOI 10.1007/978-88-470-2676-6_11, © Springer-Verlag Italia 2012

depressive symptoms according to the Centre for Epidemiological Studies' Depression Scale (CES-D), with 29 % scoring in the moderate or severe range.

11.1.1 Anxiety Disorders

Anxiety has been less investigated than depression in MS, although it is a cause of disability in these patients, too. In the literature, the scale of prevalence of anxiety disorders is very variable, between 19 % and 90 % [10–12]; this means that according to some studies, anxiety is an even more frequent disorder than depression and mood disorders in general [10, 13]. In particular, higher scores of anxiety have been reported in recently diagnosed MS patients (34 %); their partners show an even higher (40 %) incidence of anxiety [12].

Moreover, in a longitudinal 2-year study conducted in 101 recently diagnosed patients, the same group observed that MS patients and their partners continued to have high levels of anxiety and distress in the first years after diagnosis [14]. In another study conducted on 140 consecutively diagnosed MS patients, the lifetime prevalence of any anxiety disorder was 35.7 %, with generalized anxiety disorder, 18.6 %; panic disorder, 10 %; obsessive-compulsive disorder, 8.6 %, social phobia, 7.8 %; all these figures are higher than those of general population, apart from that of social phobia. Risk factors include being female, a co-morbid diagnosis of depression, and limited social support [15].

11.2 Clinical Aspects

11.2.1 Diagnostic Criteria

The Diagnostic and Statistical Manual of Mental Disorders, Fourth Edition – Text Revision (DSM-IV-TR) suggests the following features and criteria for the diagnosis of "major depressive episode":

1. Depressed mood most of the day, nearly every day, as indicated by either subjective report (e.g. feels sad or empty) or observation made by the others (e.g. appears tearful).
2. Markedly diminished interest or pleasure in all, or almost all, activities of the day, nearly every day.
3. Significant weight loss when not dieting or weight gain, or decrease or increase in appetite nearly every day.
4. Insomnia or hypersomnia nearly every day.
5. Psychomotor agitation or retardation nearly every day.
6. Fatigue or loss of energy nearly every day.
7. Feelings of worthlessness or excessive or inappropriate guilt.
8. Diminished ability to think or concentrate, or indecisiveness, nearly every day.
9. Recurrent thoughts of death, recurrent suicidal ideation without specific plan, or a suicide attempt or a specific plan for committing suicide.

Five or more of these symptoms must be contemporarily present during a 2-week period; they represent a significant change from previous functioning; the presence of at least depressed mood or loss of interest or pleasure is mandatory.

In the case of MS, these criteria raise some issues: four or five symptoms (weight changes, insomnia or hypersomnia, psychomotor agitation or retardation, fatigue and concentration problems) may be equally present as a direct consequence of the disease and not of a mood dysfunction. As will be seen in the next chapters on the etiopathogenesis of MS, a major depressive episode may be considered as the direct consequence of peculiar MRI lesions in specific parts of the brain. In this case, we need to consider MDD as a "mood disorder due to a general medical condition". Otherwise, we may hypothesize that MDD is a psychological reaction to a stress event, i.e. the consequence of an adaptation disturbance.

11.2.1.1 Anxiety Disorder

Anxiety Disorders categorize a large number of disorders where the primary feature is abnormal or inappropriate anxiety. Individuals often exhibit a variety of physical symptoms, including fatigue, fidgeting, headaches, nausea, numbness in hands and feet, muscle tension, muscle aches, difficulty swallowing, bouts of difficulty breathing, difficulty concentrating, trembling, twitching, irritability, agitation, sweating, restlessness, insomnia, hot flashes, and rashes, accompanied by an inability to fully control the anxiety.

There are several kinds of anxiety disorders:

- *Panic Disorder*: People with panic disorder have feelings of terror that strike suddenly and repeatedly with no warning. They cannot predict when an attack will occur, and many develop intense anxiety between episodes, worrying when and where the next attack will strike. This disorder may be accompanied by agoraphobia.
- *Specific phobia*: A phobia is a fear which is caused by a specific object or situation. The fear may be triggered by the actual presence of, or by the anticipation of the presence of, that object or situation.
- *Social phobia*: A persistent irrational fear of situations in which the person may be closely watched and judged by others, as in public speaking, eating, or using public facilities.
- *Obsessive Compulsive Disorder*: People with obsessive-compulsive disorder have either obsessions, or compulsions, or both. The obsessions and/or compulsions are strong enough to cause significant distress in their employment, schoolwork, or personal and social relationships.
- *Post Traumatic Stress Disorder (PTSD)*: The repeated reliving of a traumatic event.
- *Acute Stress Disorder*: a disorder following exposure to a traumatic event in which intense fear, helplessness, or horror have been experienced.
- *Generalized Anxiety Disorder*: this is much more than the normal anxiety people experience day to day. It is chronic and exaggerated worry and tension, even though nothing seems to provoke it.

Like MDD, anxiety disorders too may be the consequence of a general medical condition or substance use, including medications and drugs of abuse.

11.2.2 Clinical Presentation

From the clinical point of view, depression in MS has a distinctive presentation. The depressive mood in patients with MS has a distinctive quality, in which anxiety, irritability, anger and somatic disturbances are predominant. Apathy and anhedonia appear less frequently, as does social isolation. A longitudinal study [16] comparing different depression symptom clusters in MS showed that mood symptoms are significantly more variable over time than neuro-vegetative symptoms, which, unlike depression in neurologically healthy subjects, have a greater clinical stability. In the same study, the authors emphasized that during the development of true depressive symptoms, the inability to make use of active coping strategies over time correlates with a greater ingravescence of the clinical psychiatric picture [16].

Depression is usually more common during relapses than in remission [10, 17] and when the neurologic disability shows a progressive development; however, high prevalence has also been reported in 54 % of MS patients with a more benign course of the disease [18]. In contrast, no close relationship has been found between depressive symptoms and disease duration or degree of physical impairment in Relapsing-Remitting Multiple Sclerosis (RRMS) [10].

Finally, the risk of relapse induced by depression is an interesting topic that has been addressed in several studies, the results of which are often conflicting. In fact, while some works have reported a positive association between depression and clinical relapse of MS [10, 19, 20], with higher scores of depression in patients in relapse than in remission, other studies have found no association [21].

Recently, in a 2-year prospective study designed to determine relapse predictors in 101 RRMS patients, the relapses were not predicted either by depression or by anxiety [22]. Otherwise, the impact of depression on the clinical neurological progress of MS is still to be clarified.

The high rate of depression in MS is related to the question of suicide and self-harm gestures. Several studies have reported a higher rate of suicide in patients with MS [4, 12, 23], reaching values 7.5 times greater than in normal age-matched populations [23]. The highest risk was found in young, male patients in the first 5 years following the initial MS diagnosis [12]. Another study [4] reported a 28.6 % lifetime prevalence of suicidal intent, with 6.4 % of patients who performed self-injury gestures. However, further studies on a greater number of patients assisted in different health units, and not only in ad hoc ambulatory services for MS, are needed to clarify this aspect, considering the clinical and psychopathological complexity of the causes of suicide.

11.2.3 Depression and Interferon Treatment

The association between depression and interferon therapy has been an object of study as well as clinical interest for years. In fact, interferon-beta treatment was commonly considered a significant risk factor, inducing depressive symptoms or worsening those existing previously. This belief resulted from some isolated case reports of suicide or suicide attempts in treated patients [24, 25].

The study "Controlled High-Risk Subjects Avonex Multiple Sclerosis Prevention Study" (CHAMPS) provided more information than previous case reports; in a fairly large population of patients, a higher prevalence of depression (20 %) in interferon- treated patients than in the placebo group (13 %) was found [26]. Moreover, the longitudinal study of depression in MS conducted by Arnett and Randolph found that patients with worsening depressive symptoms were significantly more likely to have been taking interferon beta than those without mood disorders [16]. Even severe depression with suicide ideation or attempts may be observed during treatment of MS with IFN-beta [27].

On the other hand, studies apart from the CHAMPS study have not shown any correlation between interferon beta treatment and depression onset in MS patients [4, 28–30]. For example, the study "Prevention of Relapses and Disability by Interferon beta 1-a Subcutaneously in Multiple Sclerosis" (PRISMS) showed no differences in levels of depression among groups of treated patients and those in the placebo arm [29]. Finally, a study conducted by Porcel et al. [31] analyzed the emotional state of MS patients treated with interferon beta for 4 years; their data support the absence of emotional worsening with this long-term therapy. In conclusion, although it is commonly believed that interferon beta treatment could be a cause of depression, data from studies conducted on less selective populations are inconclusive, and the question is still controversial.

11.2.4 Depression and Fatigue

Depression and fatigue are extremely important facets of a patient's experience of MS and are major determinants of the quality of life. The association between depression and fatigue is a matter of great clinical interest. As a matter of fact, not only are these two symptoms often present at the same time, but it is also tempting to ascribe the presence of depression to fatigue and vice-versa, as is shown both in past and current research. Fatigue is reported by almost 75 % of MS patients [32], who describe it as an overwhelming sense of tiredness not explained by exertion, lack of energy or feeling of exhaustion that usually increases during the day. Fatigue is the symptom that patients identify as interfering most with their daily activities [33]; it may occur in any stage of MS [34], even when there is a more benign prognosis [35], and seems to be independent from disability or ambulation [34, 36, 37]. Although some studies point to positive correlations between fatigue and mood levels [32, 37], the relationship between the two symptoms has not yet been entirely clarified [38].

It is important to investigate these symptoms carefully at the time of diagnosis and throughout the course of the disease. Furthermore, the potential impact on them of immuno-modulatory treatment should be taken into account when making treatment decisions. Interestingly, a recent study reported that improvements in depression and fatigue also predicted improved accuracy in perceiving cognitive abilities from pre- to post- treatment [39]. Depression and fatigue should not be considered as secondary issues but as an integral part of disease presentation and management. As such, they should be included as endpoints of future clinical trials so that the impact of treatment on these symptoms can be better defined.

11.2.5 Depression and Cognition

It is also important to evaluate the relationship between depression and cognition in MS. Although cognitive impairment in MS patients has been recognized as an important aspect for some decades, more recent clinical research has shown how, in practice, it highly affects social and emotional functioning, the ability to perform normal daily activities, the maintenance of a satisfying occupation, and quality of life in general.

Broadly speaking, intact cognitive functioning is the result of a complex of specific cognitive abilities which function optimally together. So, both in research and in clinical practice with MS patients, it is important to identify cognitive impairments and to specify their characteristics [40].

Otherwise, cognitive impairment in MS is often a latent condition, unacknowledged by the patient and his family. Yet, in the vicious circle between cognitive and psychological impairment, the role played by each type of symptom remains unsettled. Given this, when MS patients are evaluated during cognitive tasks, it is common practice to assess other symptoms such as depression, anxiety and fatigue along with the neuropsychological situation. In fact, given the strong association between depression, fatigue, and perceived cognitive functioning, many researchers believe that perceptions of cognitive abilities are driven more by these co-morbid symptoms than by actual cognitive performance. For example, Bruce and Arnett [41] suggested that mildly depressed patients overestimate their memory impairment due to a depressive schema. Randolph, Arnett, & Freske [42] also found that depression aggravated patients' memory complaints due in part to depressive attitudes. Similarly, Middleton et al. (2006) [43] concluded that patients' perceptions of their cognitive functioning are more reflective of their emotional state and fatigue than their objective abilities. These findings are consistent with Beck's (1967) cognitive theory of depression, which states that depression is associated with a negative, pessimistic view of oneself, the environment, and the future. This negative bias can lead people to misinterpret facts and exaggerate their problems, for instance overestimating cognitive difficulties. The high frequency of cognitive changes in MS is well recognized, affecting 40–60 % of patients [44], and can have a significant impact on social and occupational functioning. Cognitive impairment can be an early feature of the demyelinating process [45–47], and may be detected in patients with clinically isolated syndromes [48] and in patients with clinically benign MS [18]. However, it is decidedly more frequent and severe in progressive forms than in relapsing-remitting ones [18]. However, the possibility of an impairment of cognitive functions as a clinical interrelation of a depressive condition can represent a substantial problem in MS patients, having a negative impact on "quality of life" and activities of daily living.

The relationship between depression and cognitive function in MS is not only a topic of clinical interest but also pertains to the development of clinical research on the disease. Altogether, most of the recent literature has found that depression in MS may really exacerbate cognitive impairment, weighing in particular on important aspects such as executive functions and working memory [49].

11.2.6 Pathogenesis of Depression in MS

A variety of mechanisms may underlie the high prevalence of depression in MS patients. Firstly, it may be understood as a reaction to stress following the diagnosis, and uncertainty about the prognosis and about the future in general. In the absence of adequate social support, and if inappropriate coping strategies are implemented, the stressful reaction may be reinforced and perpetuated. If the patient does not possess the necessary psychological resources to break this vicious circle, chronic clinical depression can develop over the course of time.

Secondly, the neuro-inflammatory nature of the MS disease process may contribute to the development or aggravation of depression. The release of pro-inflammatory cytokines such as interferon-γ, TNF–α or interleukin-6 may produce symptoms that reinforce an underlying tendency to depression, including loss of appetite, sleep disturbances, asthenia or weight loss. Moreover, biological and psychological factors may interact to exacerbate depressive symptoms: for instance, stress leads to activation of the hypothalamus-pituitary-adrenal axis and of the sympathetic nervous system, which in turn can, under certain circumstances, stimulate the immune system to release pro-inflammatory cytokines. In this context, the observation [50] that resolution of depression in MS is accompanied by a reduction in the production of pro-inflammatory cytokines is consistent with the hypothesis of a dynamic relationship between neuroinflammation and depression.

Finally, structural changes in the brain may contribute to the development of depression. The question as to whether or not depression in MS is associated with lesions in specific areas of the central nervous system has recently generated considerable interest. This question has particular relevance for understanding and treating depression in patients with MS.

A detailed report of the studies conducted on the relationships between mood disturbances and neuro-imaging in MS patients will be presented in the dedicated section of this book. The available data cannot be regarded as sufficient for answering the question of whether mood disorders in MS are the direct consequence of structural damage or if psychological processes are more important than lesions in some brain structures. The Conclusions section gives an example of an etiologic hypothesis that tries to take all the factors into account.

11.3 Clinical Evaluation

The evaluation of mood should be considered a fundamental step in the clinical assessment of an MS patient.

11.3.1 Evaluation of Depressive Symptoms

The first step involves investigating the presence of depressive symptoms, the peculiar quality of the mood, and the basic psychopathological dimensions characterizing it. We will focus particularly on the symptoms which are not so frequently noticed in depression in neurologically healthy subjects, but which are a distinctive characteristic of MS patients: irritability, anxiety, anger, somatic and neuro-vegetative symptoms. In this last case, the task of the clinician is even more arduous: to refer a somatic symptom like asthenia to a simple psychopathogical depressive correlate or, on the other hand, to a mere symptom of neurological character, taking no account of its possible psychic origin, can be risky for the clinician and harmful for the patient. There is a very fine line between the fields of psychiatry and neurology, and the boundary between them is by no means clear.

In making a proper psychopathogical evaluation aimed at establishing the presence of a clinical picture of depression, a leading role is played by the clinical conversation, which can verify the patient's subjective symptoms (sadness, anxiety, loss of pleasure or interest, reported by the subject) as well as his/her objective symptoms (flat gestures, worried look, level of personal neglect, speaking in a low-pitched monotone, presence of psychomotor retardation or, in the case of the so-called agitated forms, psychomotor agitation).

The clinical evaluation will also use a series of tests to be completed by the operator or the patient, of which the most frequently used are:

- *Beck Depression Inventory (BDI)*: a short questionnaire consisting of 21 items found through clinical observation to be representative of the symptoms and attitudes of depressed patients in the course of a clinical conversation. The fundamental assumption is that score, frequency and intensity of symptoms are directly correlated with depth of depression. The theory underlying the BDI construct is that depressive disorders are a consequence of a cognitive structuring that induces the subject to see himself and his future in a negative light.
- *State Trait Anxiety Inventory (STAI)*: the STAI is made up of two short subtests (20 items in each subtest); each one is answered on a scale of four intensity levels: the X-I subtest measures the "state anxiety" at the time that the test is administered; the X-2 subtest measures anxiety as a "trait", i.e., the tendency of the subject to produce anxious reactions under specific conditions.
- *Hamilton Depression Rating Scale (HDR-S)*: a scale probing 21 different areas which are decisive for evaluating the depressive state of the subject. The areas are: depressed mood, sense of guilt, suicidal ideation, insomnia initial, insomnia middle, insomnia delayed, work and interests, retardation of thought and speech, agitation, anxiety (psychic), anxiety somatic, gastrointestinal somatic symptoms, general somatic symptoms, genital symptoms, hypochondriasis, insight, loss of weight, diurnal variation of symptoms, depersonalization, paranoid symptoms, obsessional symptoms.

However, tools like the BDI have been questioned as a consequence of problems in the attribution of some symptoms (e.g., fatigue) to mood disturbances or to the disease itself. Accordingly, other instruments that can give separate accounts of mood and vegetative symptoms have been proposed. One of these is the Chicago Multiscale Depression Inventory, composed of three sub-scales with 14 items each to a total of 42 self-report questions; the sub-scales assess mood, vegetative symptoms and evaluation [51, 52].

In addition to using scales evaluating depression and anxiety symptoms, as mentioned before, a neurocognitive evaluation should be performed, to examine the state of the patient at baseline, and to evaluate the efficacy of treatments. In fact, the task of a psychiatric therapy, whether psychological or strictly pharmacological, will be to refine the neurocognitive performance of the subject, as well as to treat depressive symptoms. A series of scales and tests can be used for this purpose. However, this issue will be discussed in greater detail in another chapter of this volume.

11.4 Treatment

Depression in MS patients appears to be under-detected and under-treated. A recent study of 260 patients showed that as many as 26 % met the criteria for major depressive disorder (MDD); among these patients, 66 % received no antidepressant medication and 4.7 % received sub-threshold doses from their neurologist [53]. Despite the development of reliable, user-friendly screening tools to detect depression in MS patients, neuropsychiatric screening has not been widely adopted.

Treatment for depression in persons with MS should be individualized and involve psychotherapy, psychopharmacology, or a combination [54].

11.4.1 Pharmacotherapy

Although antidepressant use is common among persons with MS [55], the literature on their effectiveness is small and largely anecdotal. Some randomized trials of pharmacotherapy for depression found both desipramine and sertraline to be effective in managing depressive symptoms [56, 57]. Recently, a Cochrane review based on two randomized clinical trials showed that both desipramine and paroxetine show a trend towards efficacy in depression in MS in the short term, but both treatments were associated with adverse effects, with significantly more patients treated with paroxetine suffering from nausea or headache [58].

Several conclusions can be drawn from the available literature on pharmacotherapy for depression in MS. First, antidepressants reduce depressive symptom severity in persons with MS and should be considered for treating MDD in this patient population. However, although depressive symptoms may be responsive to pharmacotherapy, this does not necessarily result in full remission of symptoms for all individuals with MS. Research on methods for identifying MS patients or symptoms that are particularly responsive to antidepressant medications would improve the treatment of MDD among MS patients. Given the clinical characteristics of MS, the side effects of antidepressants may be particularly bothersome in this population, and lead to higher rates of non-adherence and premature treatment withdrawal in clinical practice than in clinical trials. The small sample sizes in the existing studies suggest that it is difficult to enroll large numbers of participants into pharmacotherapy clinical trials in this disease group; future research evaluating antidepressants should consider multi-center trials.

11.4.2 Psychotherapy

Several randomized controlled trials have demonstrated the effectiveness of individual cognitive behavior therapy (CBT) for treatment of major depressive disorder among persons with MS [57]. In these studies, response rates to CBT have been equal to or higher than response rates to antidepressant medications or to other psychotherapy modalities, approaching 50 % of patients. Moreover, attrition is typically very low for the CBT intervention in these trials (5 %) [57]. Recent studies have also shown that CBT delivered by telephone is an effective form of psychotherapy for depression in MS relative to usual care [59] and relative to a telephone-delivered supportive emotion-focused therapy [60]. Telephone-delivered CBT shows considerable promise as a viable and potentially cost-effective treatment for MDD that may overcome many of the common barriers to face-to-face treatment such as fatigue, stigma, and logistical issues (lack of access to treatment, transportation, child care, or financial limitations). Interpersonal therapy or behavioral activation have not been empirically evaluated as depression treatments in the MS literature. The Goldman Consensus group has recommended that psychotherapy, particularly cognitive behavior therapy (CBT), be offered as a treatment option for persons with MS and depression [54]. As skills learned through CBT produce improvements beyond the nonspecific effects of supportive treatment [57], standard CBT for depression may be considered as the treatment of choice.

11.4.3 Exercise

Although physical exercise has not been formally studied as a treatment for major depression in MS, it has widespread beneficial effects among persons with MS, including improvements in mood, pain, fatigue, quality of life, sexual functioning, recreation, and psychosocial functioning [61]. The effect of exercise on MDD has been studied in healthy adults, persons with psychiatric conditions, and older adults. Across studies, exercise appears to be more beneficial than no treatment, and in some studies it has been as effective as antidepressant medication and psychotherapy for mild to moderate depression. Exercise has also been associated with lower depression relapse rates compared to pharmacotherapy. Moderate exercise (walking 20 min a day at 60 % maximum heart rate) has been more effective than vigorous exercise and is associated with fewer drop outs [62].

References

1. Feinstein A (2004) The neuropsychiatry of multiple sclerosis. Can J Psychiatry 49:157–163
2. Patten SB, Beck CA, Williams JV, Barbui C, Metz LM (2003) Major depression in multiple sclerosis: a population-based perspective. Neurology. Neurology 9(61):1524–1527
3. Kern S, Ziemssen T (2008) Brain-immune communication. Psychoneuroimmunology of multiple sclerosis. Mult Scler 14:6–21
4. Feinstein A (2002) An examination of suicidal intent in patients with multiple sclerosis. Neurology 10(59):674–678
5. Minden SL, Schiffer RB (1990) Affective disorders in multiple sclerosis. Review and recommendations for clinical research. Arch Neurol 47:98–104
6. Minden SL, Orav J, Reich P (1987) Depression in multiple sclerosis. Gen Hosp Psychiatry 9:426–434
7. Joffe RT, Lippert GP, Gray TA, Sawa G, Horvath Z (1987) Mood disorder and multiple sclerosis. Arch Neurol 44:376–378
8. Sadovnick AD, Remick RA, Allen J, Swartz E, Yee IM, Eisen K, Farquhar R, Hashimoto SA, Hooge J, Kastrukoff LF, Morrison W, Nelson J, Oger J, Paty DW (1996) Depression and multiple sclerosis. Neurology 46:628–632
9. Chwastiak L, Ehde DM, Gibbons LE, Sullivan M, Bowen JD, Kraft GH (2002) Depressive symptoms and severity of illness in multiple sclerosis: epidemiologic study of a large community sample. Am J Psychiatry 159:1862–1868
10. Noy S, Achiron A, Gabbay U, Barak Y, Rotstein Z, Laor N, Sarova-Pinhas I (1995) A new approach to affective symptoms in relapsing-remitting multiple sclerosis. Compr Psychiatry 36:390–395
11. Stenager E, Knudsen L, Jensen K (1994) Multiple sclerosis: correlation of anxiety, physical impairment and cognitive dysfunction. Ital J Neurol Sci 15:97–101
12. Feinstein A, O'Connor P, Gray T, Feinstein K (1999) The effects of anxiety on psychiatric morbidity in patients with multiple sclerosis. Mult Scler 5:323–326
13. Stenager EN, Stenager E, Koch-Henriksen N, Brønnum-Hansen H, Hyllested K, Jensen K, Bille-Brahe U (1992) Suicide and multiple sclerosis: an epidemiological investigation. J Neurol Neurosurg Psychiatry 55:542–545
14. Janssens AC, Buljevac D, van Doorn PA, van der Meché FG, Polman CH, Passchier J, Hintzen RQ (2006) Prediction of anxiety and distress following diagnosis of multiple sclerosis: a two-year longitudinal study. Mult Scler 12:794–801
15. Korostil M, Feinstein A (2007) Anxiety disorders and their clinical correlates in multiple sclerosis patients. Mult Scler 13:67–72
16. Arnett PA, Randolph JJ (2006) Longitudinal course of depression symptoms in multiple sclerosis. J Neurol Neurosurg Psychiatry 77:606–610
17. McCabe MP (2005) Mood and self-esteem of persons with multiple sclerosis following an exacerbation. J Psychosom Res 59:161–166
18. Amato MP, Zipoli V, Portaccio E (2006) Multiple sclerosis-related cognitive changes: a review of cross-sectional and longitudinal studies. J Neurol Sci 15(245):41–46
19. Dalos NP, Rabins PV, Brooks BR, O'Donnell P (1983) Disease activity and emotional state in multiple sclerosis. Ann Neurol 13:573–577
20. Warren S, Warren KG, Cockerill R (1991) Emotional stress and coping in multiple sclerosis (MS) exacerbations. J Psychosom Res 35:37–47
21. Möller A, Wiedemann G, Rohde U, Backmund H, Sonntag A (1994) Correlates of cognitive impairment and depressive mood disorder in multiple sclerosis. Acta Psychiatr Scand 89:117–121
22. Brown RF, Tennant CC, Sharrock M, Hodgkinson S, Dunn SM, Pollard JD (2006) Relationship between stress and relapse in multiple sclerosis: Part I Important features. Mult Scler 12:453–464

23. Sadovnick AD, Eisen K, Ebers GC, Paty DW (1991) Cause of death in patients attending multiple sclerosis clinics. Neurology 41:1193–1196
24. Klapper JA (1994) Interferon beta treatment of multiple sclerosis. Neurology 44:188
25. Pandya R, Patten S (2002) Depression in multiple sclerosis associated with interferon beta-1a. Can J Psychiatry 47:686
26. Jacobs LD, Beck RW, Simon JH, Kinkel RP, Brownscheidle CM, Murray TJ, Simonian NA, Slasor PJ, Sandrock AW (2000) Intramuscular interferon beta-1a therapy initiated during a first demyelinating event in multiple sclerosis. CHAMPS Study Group. N Engl J Med 28 (343):898–904
27. Fragoso YD, Frota ER, Lopes JS, Noal JS, Giacomo MC, Gomes S, Gonçalves MV, da Gama PD, Finkelsztejn A (2010) Severe depression, suicide attempts, and ideation during the use of interferon beta by patients with multiple sclerosis. Clin Neuropharmacol 33:312–316
28. Mohr DC, Goodkin DE, Masuoka L, Dick LP, Russo D, Eckhardt J, Boudewyn AC, Bedell L (1999) Treatment adherence and patient retention in the first year of a Phase-III clinical trial for the treatment of multiple sclerosis. Mult Scler 5:192–197
29. Patten SB, Metz LM (2001) Interferon beta-1 a and depression in relapsing-remitting multiple sclerosis: an analysis of depression data from the PRISMS clinical trial. Mult Scler 7:243–248
30. Zephir H, De Seze J, Sénéchal O, Stojkovic T, Ferriby D, Deliss B, Dubus B, Verier A, Hautecoeur P, Vermersch P (2002) Treatment of progressive multiple sclerosis with cyclophosphamide. Rev Neurol (Paris) 158:65–69
31. Porcel J, Río J, Sánchez-Betancourt A, Arévalo MJ, Tintoré M, Téllez N, Borràs C, Nos C, Montalbán X (2006) Long-term emotional state of multiple sclerosis patients treated with interferon beta. Mult Scler 12:802–807
32. Ford H, Trigwell P, Johnson M (1998) The nature of fatigue in multiple sclerosis. J Psychosom Res 45:33–38
33. Kesselring J, Beer S (2005) Symptomatic therapy and neurorehabilitation in multiple sclerosis. Lancet Neurol 4:643–652
34. Krupp LB, Christodoulou C (2001) Fatigue in multiple sclerosis. Curr Neurol Neurosci Rep 1:294–298
35. Amato MP, Zipoli V, Goretti B, Portaccio E, De Caro MF, Ricchiuti L, Siracusa G, Masini M, Sorbi S, Trojano M (2006) Benign multiple sclerosis: cognitive, psychological and social aspects in a clinical cohort. J Neurol 253:1054–1059
36. Vercoulen JH, Swanink CM, Zitman FG, Vreden SG, Hoofs MP, Fennis JF, Galama JM, van der Meer JW, Bleijenberg G (1996) Randomised, double-blind, placebo-controlled study of fluoxetine in chronic fatigue syndrome. Lancet 347:858–861
37. Téllez N, Río J, Tintoré M, Nos C, Galán I, Montalban X (2006) Fatigue in multiple sclerosis persists over time: a longitudinal study. J Neurol 253:1466–1470
38. Mohr DC, Hart SL, Goldberg A (2003) Effects of treatment for depression on fatigue in multiple sclerosis. Psychosom Med 65:542–547
39. Kinsinger SW, Lattie E, Mohr DC (2010) Relationship between depression, fatigue, subjective cognitive impairment, and objective neuropsychological functioning in patients with multiple sclerosis. Neuropsychology 24:573–580
40. Chiaravalloti ND, DeLuca J (2008) Cognitive impairment in multiple sclerosis. Lancet Neurol 7:1139–1151
41. Bruce J, Arnett P (2004) Self-reported everyday memory and depression in patients with multiple sclerosis. J Clin Exp Neuropsychol 26:200–214
42. Randolph JJ, Arnett PA, Freske P (2004) Metamemory in multiple sclerosis: Exploring affective and executive contributors. Arch Clin Neuropsychol 19:259–279
43. Middleton LS, Denney DR, Lynch SG, Parmenter BA (2006) The relationship between perceived and objective cognitive functioning in multiple sclerosis. Arch Clin Neuropsychol 21:487–494
44. Rao SM, Leo GJ, Bernardin L, Unverzagt F (1991) Cognitive dysfunction in multiple sclerosis. I. Frequency, patterns, and prediction. Neurology 41:685–691

45. Achiron A, Barak Y (2006) Cognitive changes in early MS: a call for a common framework. J Neurol Sci 245:47–51
46. Jønsson A, Andresen J, Storr L, Tscherning T, Soelberg Sørensen P, Ravnborg M (2006) Cognitive impairment in newly diagnosed multiple sclerosis patients: a 4-year follow-up study. J Neurol Sci 245:77–85
47. Schulz D, Kopp B, Kunkel A, Faiss JH (2006) Cognition in the early stage of multiple sclerosis. J Neurol 253:1002–1010
48. Feuillet L, Reuter F, Audoin B, Malikova I, Barrau K, Cherif AA, Pelletier J (2007) Early cognitive impairment in patients with clinically isolated syndrome suggestive of multiple sclerosis. Mult Scler 13:124–127
49. Feinstein A (2006) Mood disorders in multiple sclerosis and the effects on cognition. J Neurol Sci 245:63–66
50. Gold SM, Irwin MR (2006) Depression and immunity: inflammation and depressive symptoms in multiple sclerosis. Neurol Clin 24:507–519
51. Nyenhuis DL, Luchetta T, Yamamoto C et al (1998) The development, standardization, and initial validation of the Chicago multiscale depression inventory. J Pers Assess 70:386–401
52. Solari A, Motta A, Mendozzi L et al (2003) Italian version of the Chicago Multiscale Depression Inventory: translation, adaptation and testing in people with multiple sclerosis. Neurol Sci 24:375–383
53. Mohr DC, Hart SL, Fonareva I, Tasch ES (2006) Treatment of depression for patients with multiple sclerosis in neurology clinics. Mult Scler 12:204–208
54. Goldman Consensus Group (2005) The Goldman Consensus statement on depression in multiple sclerosis. Mult Scler 11:328–337
55. Cetin K, Johnson KL, Ehde DM, Kuehn CM, Amtmann D, Kraft GH (2007) Antidepressant use in multiple sclerosis: epidemiologic study of a large community sample. Mult Scler 13:1046–1053
56. Schiffer RB, Wineman NM (1990) Antidepressant pharmacotherapy of depression associated with multiple sclerosis. Am J Psychiatry 147:1493–1497
57. Mohr DC, Boudewyn AC, Goodkin DE, Bostrom A, Epstein L (2001) Comparative outcomes for individual cognitive-behavior therapy, supportive-expressive group psychotherapy, and sertraline for the treatment of depression in multiple sclerosis. J Consult Clin Psychol 69:942–949
58. Koch MW, Glazenborg A, Uyttenboogaart M, Mostert J, De Keyser J (2011) Pharmacologic treatment of depression in multiple sclerosis. Cochrane Database Syst Rev 16(2):CD007295
59. Mohr DC, Likosky W, Bertagnolli A, Goodkin DE, Van Der Wende J, Dwyer P, Dick LP (2000) Telephone-administered cognitive-behavioral therapy for the treatment of depressive symptoms in multiple sclerosis. J Consult Clin Psychol 68:356–361
60. Mohr DC, Hart SL, Julian L, Catledge C, Honos-Webb L, Vella L, Tasch ET (2005) Telephone-administered psychotherapy for depression. Arch Gen Psychiatry 62:1007–1014
61. Petajan JH, Gappmaier E, White AT, Spencer MK, Mino L, Hicks RW (1996) Impact of aerobic training on fitness and quality of life in multiple sclerosis. Ann Neurol 39:432–441
62. Brown TR, Kraft GH (2005) Exercise and rehabilitation for individuals with multiple sclerosis. Phys Med Rehabil Clin N Am 16:513–555

Bipolar Disorder and Mania

12

Alberto Siracusano, Cinzia Niolu, Michele Ribolsi, and
Lucia Sacchetti

12.1 Epidemiology

It has been emphasised elsewhere in this volume that depression is more frequent in
MS patients than in the general population. The same goes for the prevalence of other
psychiatric disorders and symptoms in the general population and in MS patients, and
bipolar disorder (BD) is no exception, as has been repeatedly reported [1–3]. The BD
prevalence rate estimated in MS patients in earlier reports is higher than that shown
by more recent surveys (0.3 % to 32 % vs 0.3 % to 2.4 %) [1, 2; 4–7], the last values
being supported by an estimate in a large population study [7]. The two to three times
higher prevalence of BD in MS patients than in the general population which emerged
from previous studies remains valid, as does the recently reported prevalence of BD
in the general population of 0.24 % instead of the usually estimated 1 % [8].

Co-morbidity higher than that expected in a chance association has prompted the
search for an explanation; various factors may contribute to increasing the risk of
BD in MS patients, from genetics to lesion location, from drugs to adjustment to a
disease such as MS.

No clear evidence of a causal relationship between MS and BD has emerged
from the rarely reported cases of a manic episode as the picture of MS onset [9–12].

12.2 Clinical Picture

Neuropsychiatric symptoms are present in most MS patients; they significantly
contribute to determining a progressive deterioration in the course of the neurological
symptoms, increasing the disability level of the patient and worsening the familial,

A. Siracusano • C. Niolu • M. Ribolsi • L. Sacchetti (✉)
Psychiatric Division, University Hospital Tor Vergata, Rome, Italy
e-mail: siracusano@med.uniroma2.it; niolu@med.uniroma2.it;
mic19812000@yahoo.it; luciasa@hotmail.it

U. Nocentini et al. (eds.), *Neuropsychiatric Dysfunction in Multiple Sclerosis*,
DOI 10.1007/978-88-470-2676-6_12, © Springer-Verlag Italia 2012

social and occupational consequences of the disorder. For this reason, early recognition and effective management of the disease are an essential part of the treatment, with the goal of improving quality of life. This underlines the importance of a multidimensional approach, involving neurologists, psychiatrists, psychologists, and psychiatric rehabilitators.

A correlation has been shown between the severity of the psychiatric symptoms, the level of disability determined by the neurological picture, patient quality of life and caregiver stress [13].

BD includes a series of syndromes, characterised mainly by mood swing disorders alternating despair and euphoria.

Hippocrates was the first to systematically describe melancholia and mania. In the nineteenth century Jean-Pierre Falret defined a separate entity of mental disorder which he named "folie circulaire", characterized by a continuous cycle of depression, mania and intervals of euthymia. In the same period Baillarger defined the concept of "folie à double forme", in which mania and melancholia oscillate between each other. At the end of the nineteenth century Emil Kraepelin, the father of modern psychiatry, defined the dichotomy of "dementia praecox" and "manic-depressive disease" and made an enormous contribution to its comprehension, diagnosis and prognosis.

Currently, bipolar disorder is diagnosed according to criteria of the Diagnostic and Statistical Manual of Mental Disorders – IV edition – Text Revised – DSM-IV-TR [14], both in clinical practice and scientific research.

DSM-IV-TR classifies four different typologies of BD: BD I (characterized by the cyclic alternation of depressive and manic episodes, mixed or only with the presence of common manic episodes); BD II (characterized by the alternation of depressive and hypomanic episodes); cyclothymic disorder (characterized by a chronic alternation of low and high mood states consecutively during at least two years); and bipolar disorder not otherwise specified.

The average onset age is 18 years for BD I and 22 years for BD II.

The characteristics of a depressive episode have been described in the previous chapter. Contrary to what is observed for depression, in MS patients mania has no peculiar clinical features [15], so if we exclude neurological symptoms and signs, the psychopathology of mania is the same as in patients not affected by MS: euphoria, with an exaggerated feeling of well-being, or irritability and dysphoria, with excessive energy, logorrhea, insomnia, psychomotor agitation, impulsivity, grandiosity [11, 16].

Manic behaviour is characterized by a state of excitement that involves the cognitive as well as the affective and pulsional spheres and is manifested in mimicry, gestures, language and actions.

The DSM-IV-TR criteria for a manic episode are "a distinct period of abnormally and persistently elevated, expansive, or irritable mood, with a clear change in the usual mood of the subject, lasting at least 1 week, accompanied by hypertrophic self-esteem or grandiosity, decreased need for sleep, more talkative than usual or logorrhea, flight of ideas, distractibility, increase in goal-directed activity (either socially, at work or school, or sexually), excessive involvement in pleasurable activities that have a high potential for painful consequences". The description of diagnostic criteria contains specifications of severity (mild, moderate and severe with or without psychotic manifestations) and course (seasonal, cycling).

The need to distinguish organic affective from functional symptoms, to identify the peculiar features of the onset and progress of BD in MS patients, and to identify common etiopathogenic mechanisms that can justify the high rate of co-morbidity between the two disorders, has led to the publication of a series of case reports.

Three cases of MS with co-morbid BD, according to the DSM-IV-TR criteria, were reported by Ybarra et al. [12].

In one of these cases, the onset of BD when the patient was 18 years old and the interval before manifestations of Primary Progressive MS suggest the most likely explanation as being that of a casual association, as prevalence of MS in the reference population is 15 cases per 100,000 inhabitants, and a 1 % prevalence of BD.

In the other two cases described by Ybarra et al. [12], BD symptoms appeared years after the appearance of MS. In case 2, BD onset coincided with a deterioration in neurological condition and the shift to a secondary progressive course; the authors favoured the hypothesis that the etiology of the BD was lesional. In case 3, the separate or cumulative effect of lesions and corticosteroids was taken into consideration.

Other studies addressed the possibility that the BD was the picture of MS onset [9, 11]. Asghar-Ali et al. [11] reported two patients, presenting manic and depressive symptoms, who were diagnosed with MS without exhibiting any motor, sensory or autonomic symptoms at follow-up: the MS diagnosis was supported by radiological evidence of typical lesions. Therefore, the pathogenesis of psychiatric symptoms can be attributed to MS lesions.

12.3 Etiology

As we reported before, observations on the co-morbidity between MS and BD describe cases in which MS is diagnosed years after BD appearance, suggesting a casual association. There are also cases in which mood disorder appears as the presenting symptom of MS, or in which MS antedates BD by several years, with either a relationship with the deterioration in the neurological condition or an apparently weak relationship between neurological and affective disturbances; these cases support the hypothesis of the psycho-organic pathogenesis of BD.

As mentioned before, when we explore etiological factors in detail, it becomes evident that various options have to be considered, and in fact they have been effectively explored.

In considering the responsibility of brain lesions, it should be remembered that the brain circuits mainly involved in the control of human behaviour emerge from the prefrontal cortex (dorsolateral and orbitofrontal), the medial fronto-striatal circuits, and the temporal lobes: all these structures are richly and reciprocally connected, and are possible localization areas of MS demyelinating lesions. For instance, a lesion of the orbitofrontal circuit, responsible for empathy and inhibition of socially inappropriate behaviour, may lead to impulsivity and lability, features of manic episodes. The studies on the relationship between lesion localization and mood disorders in MS will be reported in the chapter devoted to the neuro-imaging of mood disorders in MS patients.

Another possible etiological explanation is that of a mood reaction to the pharmacological treatment of MS, and not to the disease itself, as previously suggested in some studies. Adverse behavioural effects during systemic corticosteroid treatment seem to be common, if we consider that among non-MS patients treated with corticosteroids, a meta-analysis found that severe psycho-emotional reactions occurred in nearly 6 % of patients and moderate reactions occurred in 28 % [17]. The most commonly reported corticosteroid-induced disorders include mania, depression or mixed states [18].

Patients receiving short-term corticosteroid therapy tend to present with euphoria or hypomania, while long therapy tends to induce depressive symptoms [19].

Among MS patients treated with corticosteroids or ACTH, two systematic studies found that 40 % of patients became depressed, 31 % were hypomanic, 11 % experienced both conditions and 16 % were psychotic [20, 21].

ACTH appeared to be more potent than prednisone in inducing hypomania or mania.

While there are some reports on the subject, the role of other drugs like muscle relaxants or interferons are still to be more adequately investigated.

Although research on the etio-pathogenesis of the association between BD and MS is still limited, studies have also focused on a common genetic susceptibility for both conditions. The results on the contribution of genetic vulnerability in the onset of BD in MS patients are poor and conflicting [22, 23].

The HLA system, which controls the immunity response, has been implicated in the vulnerability for many diseases with an autoimmune component. Many studies have reported an increased frequency of certain HLA antigens in MS patients [24, 25].

Schiffer at al. [22] were the first to explore the relationship between MS and affective disorders from a genetic point of view. They studied 56 patients and noticed an increased frequency of HLA-DR5 and a decreased frequency of HLA-DR1 and HLA-DR4 in bipolar MS patients with a family history of affective disorder. A possible role of HLA genes in the connection between MS and BD is also suggested by the results of other studies [26, 27].

Although the results need further investigation and validation of factors correlating MS and BD, these studies support the hypothesis that genes very close to the HLA region on chromosome 6 may constitute one of the elements in the multi-factorial etiology of BD and MS. More recent studies have focused on genetic polymorphisms related to the serotonin transporter gene as hypothetical susceptibility loci for both conditions [28, 29].

12.4 Diagnosis

The diagnosis of mania is based on the presence of signs and symptoms which respond to the criteria of manic or hypomanic episodes according to the DSM-IV-TR.

The Mania Rating Scale (MRS) is one of the psychometric tools with which we can obtain objective and sharable information about the clinical picture of patients and monitor the course of the symptoms [30].

This is an 11-item scale that explores the key symptoms of mania (mood, motor activity, quantitative and formal thought disorders, critical capacity, aggressiveness, libido, sleep and general attitude). Assessment of severity is made on the basis of the patient's report of his own condition within the past 48 h, and of behavioural observation by the clinician during the interview (with a relative priority for the latter). The scale must be used only as a tool for the quantitative assessment of mania and not for diagnosis. It was created specifically for the assessment of manic symptoms and their shifts with treatment, and is therefore expected to be used at least in pre- and post-treatment and in intermediate intervals of time.

Another issue is the problem of differential diagnosis in bipolar disorders.

First of all, it is important to distinguish hypomania from mania by the absence in hypomania of psychotic features, by its lower symptomatic severity, by the duration of an episode, by the non-significant reduction of the subject's social and working functions connected to psychiatric symptoms.

Moreover, the diagnostic boundaries between mixed states, mania and depression are not clear, and may represent the most complex aspect in the diagnosis of bipolar disorder. Observation of the longitudinal course of the disease is fundamental; useful factors in orienting the diagnosis toward a clinical picture of bipolar disease are the presence of mood elevation, familiarity for affective disorders, and a lower presence of negative symptoms characteristic of psychotic disorders. Short psychotic episodes could be a clinical manifestation of MS [31]. In clinical and MRI studies, Feinstein et al. [32] found that psychotic symptoms were usually transient, developed later than MS onset and appeared at a later age in MS patients than in non-MS individuals. Psychotic patients usually present larger lesions around the temporal horn of the lateral ventricles.

A further differentiation concerns borderline personality disorder, which presents periods of mood alteration, mainly emotional lability and irritability. Early onset and a persistent clinical picture are two elements which help distinguish personality disorder from Axis I disorder.

In the case of MS patients, the organic origin of mood elevation may be confirmed by a later onset age and by the absence of familiarity for affective disorders, as well as a temporal correlation between the two disorders.

There are two clinical conditions found in MS patients that we have to distinguish from bipolar mania: euphoria, which is actually thought to be a possible profound personality change related to extensive cerebral damage and severe cognitive impairments [33], and pseudo-bulbar affect (PBA), characterized by involuntary and inappropriate outbursts of laughter and/or crying, unconnected to the patient's affective state. Euphoria and PBA will be treated in another section of this book [34, 35].

In addition, in literature some cases are described in which the prevailing psychiatric symptoms are characterized by changes in behaviour such as confabulations, paranoid ideation, belligerence, hypersexuality, Klüver-Bucy syndrome and conduct of abuse which must be distinguished from possible manic symptoms.

At the end of this section devoted to BD diagnosis, we would like to emphasise that following cases of manic episodes heralding MS, the latter disease should be included in the diagnostic workout in BD cases.

12.5   Treatment

First of all, the treatment of psychiatric disorders in MS must involve a specific integrated approach for each single patient. The first step in patient management means communicating the diagnosis, and informing the patient about the mechanisms of action, the efficacy and side effects of pharmacological treatments, the course and prognosis of the disorder; awareness of the existence of symptomatic therapies which can have an effect on the psychological aspects of the disease is of prime importance.

Therapeutic strategies for BD treatment include both pharmacological and psychological therapy.

At present, there are no guidelines for the treatment of manic episodes in MS patients. Published studies suggest the use of mood stabilizers (carbolithium, valproic acid, carbamazepine), antipsychotics and benzodiazepines [36, 37]; they do not describe any major pharmacological interaction with interferon, the main therapy used in treating MS.

According to the guidelines of the American Psychiatric Association (APA), first-line pharmacological treatment is based on lithium or valproic acid, in monotherapy or associated with an antipsychotic, possibly atypical (such as olanzapine or risperidone) for a better side effect profile; second-line mood stabilizers are carbamazepine and oxcarbazepine. Short-term treatment with benzodiazepines, clonazepam as first-line choice, may be useful mainly in the initial phase of therapy or in the phases of psychomotor excitement.

Psychotherapy can be used along with the pharmacological treatment but, unlike the situation for depression, no data exist on the efficacy of any psychotherapeutic approach for BD.

In cases of mania induced by steroids, prophylactic treatment with carbolithium and the progressive reduction of corticosteroids is indicated [38].

Despite the fact that the coexistence of psychopathological symptoms in MS is well established, most affected patients are still under-diagnosed and undertreated. A shift involving the clinical suspicion and treatment of psychiatric disorders in MS seems to be fundamental, as these disorders can have a considerable impact on the patients' quality of life and also on their adherence to MS treatment. On the other hand, a lack of response to the standard treatment of patients with a supposed primary BD should generate suspicion of an "organic" etiology, of which MS is one possibility [39].

References

1. Schiffer RB, Wineman NM, Weikamp LR (1986) Association between bipolar affective disorder and multiple sclerosis. Am J Psychiatry 143:94–95
2. Joffe RT, Lippert GP, Gray TA, Sawa G, Horvath Z (1987) Mood disorder and multiple sclerosis. Arch Neurol 44:376–378
3. Lyoo IK, Seol HY, Byun HS, Renshaw PF (1996) Unsuspected multiple sclerosis in patients with psychiatric disorders: a magnetic resonance imaging study. J Neuropsychiatry Clin Neurosci 8(1):54–59
4. Minden SL, Schiffer RB (1990) Affective disorders in multiple sclerosis. Review and recommendations for clinical research. Arch Neurol 47:98–104
5. Diaz-Olavarrieta C, Cummings JL, Velazquez J, Garcia de la Cadena C (1999) Neuropsychiatric manifestations of multiple sclerosis. J Neuropsychiatry Clin Neurosci 11:51–57
6. Edwards LJ, Costantinescu CS (2004) A prospective study of conditions associated with multiplks sclerosis in a cohort of 658 consecutive outpatients attending a multiple sclerosis clinic. Mult Scler 10:575–585
7. Marrie RS, Horwitz R, Cutter G et al (2009) The burden of medical comorbidity in multiple sclerosis: frequent, underdiagnosed and undertreated. Mult Scler 15:385–392
8. Perala J, Suvisaari J, Saami SI et al (2007) Lifetime prevalence of psychotic and bipolar I disorders in a general population. Arch Gen Psychiatry 64:19–28
9. Hutchinson M, Starck J, Buckley P (1993) Bipolar affective disorder prior to the onset of multiple sclerosis. Acta Neurol Scand 88:388–393
10. Casanova MF, Kruesi M, Mannheim G (1996) Multiple sclerosis and bipolar disorder: a case report with autopsy findings. J Neuropsychiatry Clin Neurosci 8:206–208
11. Ali-Asghar AA, Taber KH, Hurley RA, Hayman LA (2004) Pure neuropsychiatric presentation of multiple sclerosis. Am J Psychiatry 161:226–231
12. Ybarra MI, Moreira MA, Reis Araújo C, Lana-Peixoto MA, Teixeira AL (2007) Bipolar disorder and multiple sclerosis. Arq Neuropsiquiatr 65(4B):1177–1180
13. Figved N, Myhr KM, Larsen JP, Aarsland D (2007) Caregiver burden in multiple sclerosis: the impact of neuropsychiatric symptoms. J Neurol Neurosurg Psychiatry 78:1097–1102
14. American Psychiatric Association (2000) The diagnostic and statistical manual, 4th. American Psychiatric Press, Washington, DC
15. Isaksson AK, Ahlstrom G (2006) From symptom to diagnosis: illness experiences of multiple sclerosis patients. J Neurosci Nurs 38:229–237
16. Jongen PJ (2006) Psychiatric onset of multiple sclerosis. J Neurol Sci 245:59–62
17. Lewis DA, Smith RE (1983) Steroid-induced psychiatric syndromes. A report of 14 cases and a review of the literature. J Affect Disord 4:319–332
18. Warrington TP, Bostwick JM (2006) Psychiatric adverse effects of corticosteroids. Mayo Clin Proc 8:1361–1367
19. Bolanos SH, Khan DA, Hanczyc M, Bauer MS, Dhanani N, Brown ES (2004) Assessment of mood states in patients receiving long-term corticosteroid therapy and in controls with patient-rated and clinician-rated scales. Ann Allergy Asthma Immunol 92:500–505
20. Ling MHM, Perry PJ, Tsuang MT (1981) Side effects of corticosteroid therapy. Arch Gen Psychiatry 38:471–477
21. Minden SL, Orav J, Schildkraut JJ (1988) Hypomanic reactions to ACTH and prednisone treatment for multiple sclerosis. Neurology 38:1631–1634
22. Schiffer RB, Weitkamp LR, Wineman NM, Gufformsen S (1988) Multiple sclerosis and affective disorder: family history, sex and HLA-DR antigens. Arch Neurol 45:1345–1348
23. Joffe RT, Lippert GP, Gray TA (1987) Personal and family history of affective illness in patients with multiple sclerosis. J Affect Disord 12:63–65
24. Batchelor JR, Compston A, McDonald WI (1987) The significance of the association between HLA and multiple sclerosis. Br Med Bull 34:279–284

25. Ebers GC, Paty DW, Stiller CR et al (1983) HLA-typing in multiple sclerosis sibling pairs. Lancet 2:88–90
26. Modrego PJ, Ferrandez J (2000) Familial multiple sclerosis with repetitive relapses of manic psychosis in two patients (mother and daughter). Behav Neurol 12:175–179
27. Bozikas VP, Anagnostouli MC, Petrikis P et al (2003) Familial bipolar disorder and multiple sclerosis: a three-generation HLA family study. Prog Neuropsychopharmacol Biol Psychiatry 27:835–839
28. Oliveira JR, Carvalho DR, Pontual D et al (2000) Analysis of the serotonin transporter polymorphism (5-HTTLPR) in Brazilian patients affected by dysthymia, major depression and bipolar disorder. Mol Psychiatry 5(4):348–349
29. Mansour HA, Talkowski ME, Wood J et al (2005) Serotonin gene polymorphisms and bipolar I disorder: focus on the serotonin transporter. Ann Med 8:590–602
30. Young RC, Biggs JT, Ziegler VE, Meyer DA (1978) A rating scale for mania: reliability, validity and sensitivity. Br J Psychiatry 133:429–435
31. Davidson K, Bagley CR (1969) Schizophrenia-like psychoses associated with organic disorders of the central nervous system: a review of the literature. Br J Psychiatry 4:113–184
32. Feinstein A, du Boulay G, Ron MA (1993) Psychotic illness in multiple sclerosis: a clinical and MRI study. Br J Psychiatry 161:680–685
33. Rabins PV (1990) Euphoria in multiple sclerosis. In: Rao S (ed) Neurobehavioral aspects of multiple sclerosis. Oxford University Press, New York, pp 180–185
34. Feinstein A, Feinstein K, Gray T, O'Connor P (1997) Prevalence and neurobehavioral correlates of pathological laughing and crying in multiple sclerosis. Arch Neurol 54:1116–1121
35. Feinstein A, O'Connor P, Gray T, Feinstein K (1999) Pathological laughing and crying in multiple sclerosis: a preliminary report suggesting a role or the prefrontal cortex. Mult Scler 5:69–73
36. Krupp LB, Rizvi SA (2002) Symptomatic therapy for under-recognized manifestations of multiple sclerosis. Neurology 58(Suppl 4):S32–S39
37. Ameis SH, Feinsein A (2006) Treatment of neuropsychiatric conditions associated with multiple sclerosis. Expert Rev Neurother 6:1555–1567
38. Falk WE, Mahnke MW, Poskanzer DC (1979) Lithium prophylaxis of corticotropin-induced psychosis. JAMA 241:1011–1012
39. Iacovides A, Andreoulakis E (2011) Bipolar disorder and resembling special psychopathological manifestations in multiple sclerosis: a review. Curr Opin Psychiatry 4:336–340

Mood Dysfunctions in MS and Neuroimaging

13

Antonio Gallo, Rosaria Sacco, and Gioacchino Tedeschi

Mood dysfunctions (MD) in Multiple Sclerosis (MS) patients have long been considered as reactive disorders. More recently, however, a number of neuroimaging studies have suggested a neurobiological basis for MD.

In one of the first investigations, *Rabins* et al. [1] examined the relationship between brain computed tomography (CT) findings and MD in patients with MS and spinal cord injuries (SCI). The authors showed that depression in MS patients is at least partially determined by the presence of brain involvement since: (i) average depression scores were similar in MS and SCI patients, but MS patients with brain involvement were more depressed than those with spinal cord-limited disease; (ii) depression scores were unrelated to functional disability but correlated with the degree of neurological impairment.

Further support for a neurobiological basis for MD in MS comes from Magnetic Resonance Imaging (MRI) studies, which show an association between spatial distribution of T2 lesions and psychiatric or depressive symptoms in MS. In 1987, *Honer* et al. [2] found a significant correlation between T2 lesion load (LL) in the temporal lobes and psychiatric symptoms in a small group of MS patients. Following studies [3–5] confirmed and expanded these preliminary results, showing a relationship between signs/symptoms of depression and involvement of the frontal and temporal lobes of the right hemisphere. *Zorzon* et al. [3], in particular, reported diagnosis of major depression in 19 % of a MS sample (n = 95) using the Hamilton Rating Scale for Depression (HAM-D) and a psychiatric interview by a clinician blind to MRI findings. After this evaluation, regional brain volumes and the T2 and T1-LL of the depressed and non-depressed patients were compared. The authors reported that severity of depression as well as a diagnosis of major depression correlated with right frontal T2 and T1-LL as well as right temporal lobe

A. Gallo (✉) • R. Sacco • G. Tedeschi
Istituto di Scienze Neurologiche, Seconda Università di Napoli, Naples, Italy
e-mail: antonio.gallo@unina2.it; s.rosaria@libero.it; gioacchino.tedeschi@unina2.it

U. Nocentini et al. (eds.), *Neuropsychiatric Dysfunction in Multiple Sclerosis*,
DOI 10.1007/978-88-470-2676-6_13, © Springer-Verlag Italia 2012

volume. They also noted that severity of depressive symptoms correlated significantly with volumes of the temporal lobe and right hemisphere. Subsequently, *Di Legge* et al. [5] investigated longitudinally a group of patients with clinically isolated syndromes (CIS) and tested the relationship between emotional changes, brain lesion burden and development of MS. They found that: (i) CIS patients were more prone to symptoms of anxiety and depression than healthy controls (HC); (ii) lesion burden in the right temporal region positively correlated with severity of depressive symptoms, confirming this region as a key area for developing depressive symptoms, even at the earliest stages of the disease.

Interestingly, the results of the above-mentioned studies agree and complement those obtained in an earlier work conducted by *Pujol* et al.[6] which showed that the damage of the white matter (WM) bundle connecting frontal and temporal lobes (i.e., the arcuate fasciculus) was associated with the severity of the depressive symptoms measured by the Beck Depression Inventory (BDI). It should be noted, however, that only a few patients had scores in the moderate to severe range of BDI. Thus it is fairly uncertain what the implications of these results might be for MS patients diagnosed as having major depression.

In another study [7], the investigators explored the association between the presence of affective disorders and (i) MRI disease activity as measured by the number of post-contrast active lesions, or (ii) hypothalamic-pituitary-adrenal axis (HPA) dysfunction as assessed by a corticotropin-releasing hormone (CRH) stimulation test after dexamethasone suppression. Compared with controls, MS patients had higher scores on depression and anxiety scales and exhibited a lack of suppression of cortisol release after dexamethasone pretreatment. Both affective symptoms and neuroendocrine abnormalities were correlated with the presence of contrast-enhancing lesions on MRI; however, no association with the degree of neurologic impairment was observed. Affective and neuroendocrine disorders were related to inflammatory disease activity but not to the degree of disability, thus supporting the hypothesis that MD might be causally associated with brain injury.

Another important contribution to the topic comes from *Bakshi* et al. [8], who studied a group of MS patients whose depressive symptoms were assessed by two multiple choice questionnaires, the HAM-D and the BDI. MRI parameters selected for the analysis were T2- and T1-LL as well as regional brain atrophy. After correcting for EDSS, the presence of depression was predicted by superior frontal and superior parietal hypointense T1 lesions (the so-called black holes), while the severity of depression was predicted by (i) superior frontal, superior parietal and temporal T1 lesions, (ii) third and lateral ventricles enlargement, and (iii) frontal atrophy. Contrariwise, depressive scores were not related to T2 lesions or the presence of contrast-enhancing lesions. The authors concluded that atrophy and cortical-subcortical disconnection due to frontal and parietal WM destructive lesions might contribute to depression in MS.

All the above-mentioned hypothesis finds support in a number of brain functional-metabolic studies (i.e. PET and SPECT) showing a functional disconnection of cortical associative areas, and particularly at the level the frontal lobes, in patients with primary and secondary depression [9–12]. To this regard, one of the most relevant -metabolic studies was conducted by *Sabatini* et al. [13] using MRI and SPECT to investigate the relationship between depression and both structural and functional brain abnormalities in MS patients. A comparison between depressed and non-depressed subjects showed no significant differences in the number, side, location and area of demyelinating lesions; contrariwise, regional cerebral blood flow asymmetries in the limbic cortex distinguished the two groups. An analysis of variance confirmed a significant effect of depression on the perfusion asymmetries in the limbic cortex. Finally, perfusion asymmetries in the limbic cortex significantly correlated with severity of depression. Based on their results, the authors suggested that depression in MS patients might be induced by a disconnection between subcortical and cortical areas involved in the function of the limbic system. Taken together, these and other data [8, 13] suggest that mood disorders in MS might be strictly linked to disruption of frontal and limbic serotoninergic circuits.

More recently, advanced MRI techniques have further contributed to our understanding of the pathophysiology of depression in MS. In a study by *Feinstein* et al. [14], the authors tested the association between major depression and structural brain abnormalities in MS patients using a combined approach. Together with the extimation of the regional T2/T1-LL, indeed, the brain was segmented into 13 pre-specified sub-regions where the relative proportion of gray matter (GM), WM, and cerebrospinal fluid (CSF) was computed. Compared with the euthymic subjects, those with major depression had a greater T2/T1-LL in the left medial inferior prefrontal cortex. There was also evidence of a lower GM volume and a larger CSF volume in the left anterior temporal region. The final analysis showed that T2-LL in the left medial inferior prefrontal cortex as well as CSF volume in the left anterior temporal cortex were independent predictors of depression, suggesting that both lesion burden and atrophy contribute to MD in MS. It is also worthy of note that in this study the diagnosis of major depression came from a structured psychiatric interview evaluating DSM-IV criteria. A subsequent work by the same group of researchers [15] combined the above-mentioned MRI segmentation methodology with diffusion tensor imaging (DTI) to further characterize both WM lesions and normal appearing brain tissue (NABT) in depressed MS subjects. In particular, the average fractional anisotropy (FA) and mean diffusivity (MDif) of Normal Appearing WM (NAWM) and Normal Appearing GM (NAGM) were computed for each segmented brain region. Depressed subjects had a higher T1-LL in the right medial inferior frontal region, a smaller NAWM volume in the left superior frontal region, and lower NAWM-FA and higher NAGM-MDif values in the left anterior temporal regions. Depressed subjects also had higher MDif in right inferior frontal T2 lesions. The conclusions were that the presence of DTI abnormalities in the NAWM and NAGM of depressed subjects highlights the usefulness of more subtle measures of structural brain damage in the study of MD in MS.

Functional MRI (fMRI) is a non-conventional MRI technique able to investigate the activity of cortical areas during specific tasks or resting conditions. Relying also on its ability to explore the functional connectivity of brain areas modulating affective behaviour (e.g. the prefrontal cortex (PFC) and amygdala), fMRI appears a very promising tool to investigate the neurobiology of emotions.

Thus far, fMRI studies have explored the neural substrates of emotion processing more than the specific functional correlates of MD in MS [16–18]. Preliminary findings offer new insights into the neurobiological mechanisms of emotions in MS and provide evidence that they resemble those described in other psychiatric disorders [19, 20].

In conclusion, conventional and non conventional MRI studies conducted so far strongly support a biological basis for MD in MS. In particular, there is growing evidence to suggest an association between the presence/severity of depression and a greater involvement of the fronto-temporal regions. Future MRI studies involving larger samples of subjects and advanced methodologies will have to further clarify the contribution of neuroimaging techniques to the study, evaluation and management of mood disorders in MS.

References

1. Rabins PV, Brooks BR, O'Donnell P et al (1986) Structural brain correlates of emotional disorders in multiple sclerosis. Brain 109:585–597
2. Honer WG, Hurwitz T, Li DK, Palmer M, Paty DW (1987) Temporal lobe involvement in multiple sclerosis patients with psychiatric disorders. Arch Neurol 44(2):187–90
3. Zorzon M, de Masi R, Nasuelli D et al (2001) Depression and anxiety in multiple sclerosis. A clinical and MRI study in 95 subjects. J Neurol 248(5):416–21
4. Zorzon M, Zivadinov R, Nasuelli D et al (2002) Depressive symptoms and MRI changes in multiple sclerosis. Eur J Neurol 9(5):491–6
5. Di Legge S, Piattella MC, Pozzilli C et al (2003) Longitudinal evaluation of depression and anxiety in patients with clinically isolated syndrome at high risk of developing early multiple sclerosis. Mult Scler 9(3):302–6
6. Pujol J, Bello J, Deus J et al (1997) Lesions in the left arcuate fasciculus region and depressive symptoms in multiple sclerosis. Neurology 49(4):1105–10
7. Fassbender K, Schmidt R, Mössner R et al (1998) Mood disorders and dysfunction of the hypothalamic-pituitary-adrenal axis in multiple sclerosis: association with cerebral inflammation. Arch Neurol 55(1):66–72
8. Bakshi R, Czarnecki D, Shaikh ZA et al (2000) Brain MRI lesions and atrophy are related to depression in multiple sclerosis. Neuroreport 11(6):1153–1158
9. Drevets WC, Price JL, Simpson JR Jr et al (1997) Subgenual prefrontal cortex abnormalities in mood disorders. Nature 386(6627):824–7
10. Biver F, Goldman S, Delvenne V et al (1994) Frontal and parietal metabolic disturbances in unipolar depression. Biol Psychiatry 36(6):381–8
11. Austin MP, Dougall N, Ross M et al (1992) Single photon emission tomography with 99mTc-exametazime in major depression and the pattern of brain activity underlying the psychotic/neurotic continuum. J Affect Disord 26(1):31–43
12. Mayberg HS, Starkstein SE, Sadzot B et al (1990) Selective hypometabolism in the inferior frontal lobe in depressed patients with Parkinson's disease. Ann Neurol 28(1):57–64
13. Sabatini U, Pozzilli C, Pantano P et al (1996) Involvement of the limbic system in multiple sclerosis patients with depressive disorders. Biol Psychiatry 39(11):970–5
14. Feinstein A, Roy P, Lobaugh N et al (2004) Structural brain abnormalities in multiple sclerosis patients with major depression. Neurology 62(4):586–90
15. Feinstein A, O'Connor P, Akbar N et al (2010) Diffusion tensor imaging abnormalities in depressed multiple sclerosis patients. Mult Scler 16(2):189–96

16. Passamonti L, Cerasa A, Liguori M et al (2009) Neurobiological mechanisms underlying emotional processing in relapsing-remitting multiple sclerosis. Brain 132:3380–91
17. Jehna M, Langkammer C, Wallner-Blazek M et al (2011) Cognitively preserved MS patients demonstrate functional differences in processing neutral and emotional faces. Brain Imaging Behav 5(4):241–51
18. Krause M, Wendt J, Dressel A et al (2009) Prefrontal function associated with impaired emotion recognition in patients with multiple sclerosis. Behav Brain Res 205(1):280–5
19. DeRubeis RJ, Siegle GJ, Hollon SD (2008) Cognitive therapy versus medication for depression: treatment outcomes and neural mechanisms. Nat Rev Neurosci 9(10):788–96, Review
20. Drevets WC, Videen TO, Price JL, Preskorn SH, Carmichael ST, Raichle ME (1992) A functional anatomical study of unipolar depression. J Neurosci 12(9):3628–41

Psychosis

14

Patrizia Montella, Manuela de Stefano, Daniela Buonanno,
and Gioacchino Tedeschi

The reported occurrence of psychiatric disorders along with motor and sensorial symptoms in Multiple Sclerosis (MS) dates back to the first case reports of the nineteenth century. In 1926 Cottrell and Wilson divided "mental symptomatology" of MS into intellectual, emotional and psychiatric disorders [1]. Although the correlation between somatic symptoms and affective disorders such as depression and anxiety is frequently reported, few studies have examined psychotic symptoms in MS patients. These include behavioral and perceptive disorders, thinking incoherence, signs of depersonalization and de-realization.

In the last few years, advanced neuro-imaging techniques which have implemented our knowledge on the mind-body duality, have supported a growing interest in mental and psychic functioning. With regard to MS, this means that the "classical" interpretation of psychiatric symptoms as a reaction to feelings connected to the diagnosis and clinical course of the disease has been overtaken by a different perspective: affective, perceptive, behavioral and psychotic symptoms could be related to focal and/or diffuse central nervous system (CNS) damage.

The association between MS and psychiatric symptoms is now also interpreted as a possible "disease effect". Recent studies using Magnetic Resonance (MR) and Diffusion Tensor Imaging (DTI) [2] hypothesize that, in MS patients, cortico-subcortical disconnection and involvement of the limbic system could determine the occurrence of authentic 'psychiatric relapses' [3]. A similar process has also been hypothesised for schizophrenia. In addition, Miller et al. [4] and Clerici et al. [5] suggested that both MS and schizophrenic patients have similar profiles of cytokine levels, such as interleukin-1-b, interleukin-6, interleukin-12, interferon-c and tumor necrosis factor-α, which could also contribute to the development of psychosis.

P. Montella (✉) • M. de Stefano (✉) • D. Buonanno (✉) • G. Tedeschi (✉)
Istituto di Scienze Neurologiche, Seconda Università di Napoli, Naples, Italy
e-mail: patrizia.montella@unina2.it; manueladestefano@hotmail.it; danbuonanno@virgilio.it; gioacchino.tedeschi@unina2.it

U. Nocentini et al. (eds.), *Neuropsychiatric Dysfunction in Multiple Sclerosis*,
DOI 10.1007/978-88-470-2676-6_14, © Springer-Verlag Italia 2012

However, clinical practice shows that psychiatric symptomatology could represent the onset symptom of MS as well as an adverse effect of specific therapies [3, 6–8].

Moreover, case reports of patients with a history of psychiatric hospitalization reveal clinical and neuroimaging data compatible with MS in 0.3–0.8 % of cases [9, 10].

Correlation between MS and psychiatric symptoms should be consequently studied for making correct diagnoses and for tailoring individual treatment strategies.

14.1   Epidemiology

The low occurrence of psychotic disorders in MS and the lack of studies on their evolution bring to insufficient epidemiological data for definite conclusions.

Several case reports and clinical studies describe the occurrence of delusional disorders, affective psychoses and delirium, but the prohibitive sample requirements and the heterogeneity of diagnostic methods and codes used in clinical practice determine the lack of population-based studies and the uncorrect evaluation of this clinical phenomenon.

To address these problems, Surridge et al. [11] assessed the neuropsychiatric symptoms of 108 MS patients, identifying depression (27 % of cases), euphoria (26 % of cases), personality changes, irritability (40 % of cases) and psychotic symptoms (4 % of cases). Skegg et al. [12] studied a population of 12,000 persons in New Zealand, finding 91 MS patients. Among them, 16 % had been referred to a psychiatrist before the neurological diagnosis, and for half of these patients the psychiatric diagnosis was the only one for a long period. Stenager and Jansen reported that 12 % of 336 MS patients had required psychiatric admission, 2 % of them before the neurological diagnosis [13].

Patten et al. [14] provided the first epidemiologic evidence of an association between MS and psychotic disorders. This longitudinal study, covering the period 1985–2003, was conducted in Alberta (Canada), where all citizens refer to the same health care system, which uses only one diagnostic code (International Classification of Disease, 9th edition, ICD-9 [15]). In the study sample of 2.45 million citizens (mean age 42.4 years), the estimated MS prevalence ranged from 0.3 to 0.5 %. Using ICD-9 codes 295 (schizophrenia-spectrum disorders), 297 (delusional disorders), 292, 293, 294 (drug-induced, transient psychotic disorders), the prevalence of psychotic disorders in MS patients was 2–3 %, there was no sex prevalence, and the peak of maximal incidence (4 %) was in the 15–24 year age group. In the general population the prevalence of psychosis was 0.5–1 % and its increase was age-related.

In Italy, Lo Fermo et al. [16] retrospectively evaluated medical records of 682 patients with defined MS diagnosis referring to the University of Catania's MS center from 1997 to 2007. They found that 16 patients (2.3 %) had a psychiatric onset and 5 of them had isolated psychotic symptoms.

The real frequency of psychiatric onset of MS [3, 6–8] is probably still underestimated because of the great heterogeneity of literature on this topic. Furthermore, the emotional impact of florid psychotic symptomatology and the low propensity of general practitioners, psychologists and psychiatrists to consider a neurological diagnosis for psychotic patients with no or "soft" neurological signs could interfere with the correct evaluation of the problem, leading to the wide variability of results [17].

In line with this topic, Ashgar-Alì et al. [18] proposed a diagnostic hypothesis of MS, even without neurological abnormalities, for patients with acute psychosis but no previous psychiatric history, and in presence of late onset, of previous symptoms or signs compatible with CNS dysfunction, and of a limited response to conventional psychotropic medication. Finally, the increasing frequency of case-reports describing patients with psychotic onset highlight the importance of an exhaustive clinical anamnesis, as previous psychiatric symptoms can be the key for the correct diagnosis, which can thus be made retrospectively [3].

14.2 Clinical Aspects

Psychotic disturbances can be related to focal and/or diffuse CNS damage and can occur as isolated symptoms or in association with somatic complaints, as the first signs or during the course of MS [19]. Psychotic disturbances may also be an adverse effect of specific therapies [20].

In MS patients, a pre-morbid psychological profile in relation to the development of psychosis has not yet been fully described, though it is evident that different structures of personality determine the use of different coping strategies for managing disease-related stress. Nonetheless, it is extremely important to consider the influence and the predictive meaning of social and environmental factors such as family structure, years of education, genetic predisposition, constitutional characteristics and job position.

Due to their symptomatic nature, the clinical features of psychotic disorders in MS are quite different from those ones of schizophrenia: age of onset is more variable, responses to therapies are more immediate and their evolution is unpredictable.

Taken together, the current data describe clinical courses of psychotic symptoms similar to those of MS. Matthews et al. [21] described two patients presenting an acute psychiatric problem with subsequent complete remission. Conversely, Lo Fermo et al. [16] and Blanc et al. [3] describe a long-term persistence of psychotic disorders, particularly in patients with psychotic onset, suggesting the need for long-term treatment. Moreover, retrospective studies have found no correlation between psychotic onset, clinical course of the disease (RR or SP) and physical disability of patients [16].

The most common MS-related psychiatric symptoms, at onset as well as during the disease, are the positive ones, such as grandiosity [18], persecution mania [22], jealousy [18], erotomania [23], delusions of guilt and ruin [3], visual and auditory hallucinations [24], loosening of associations [18, 25, 26], incoherence and tangentiality of speech [17, 18, 21], depersonalisation, derealisation [17], inappropriate behavior [21], hyper-religiosity [17, 18], hyper-sexuality [18], and perceptive disorders [17, 22, 25].

Negative symptoms, such as avolition, affective flattering, anhedonia and social withdrawal, catatonia, inertia and apathy, are less frequent [3, 21, 27].

Descriptions of major depressive symptoms with melancholic, psychotic [3, 28] and catatonic features are very rare and more frequent at disease onset [3, 29–31].

Delusions [24], maniacal syndrome and major depression [14] are frequently reported during the course of high-dose or long-term therapies with corticosteroids and immunotherapies with interferon [18]; patients report good response to neuroleptics and do not need to stop current therapy.

Delusions, dysphoria, depersonalization and hallucinations are frequently reported as adverse effects of cannabinoids used for managing pain and spasticity [23].

Finally, the clinical course and the possibility that psychiatric symptoms may be the onset signs of MS highlight the need to include the disease in the differential diagnosis with psychiatric pathologies [21].

14.3   Assessment

Because of the lack of multidimensional tools for neurologic disorders with psychotic manifestations, diagnosing psychosis in MS patients requires a combination of several clinical methods, such as the patient's clinical history, neurological examination and psychodiagnostic evaluation. This makes difficult to compare current studies in literature.

To define the neuropsychiatric dimensions of MS, Diaz-Olavarrieta et al. [22] used the Neuropsychiatric Inventory (NPI) [32], a caregiver-based rating scale of established validity and reliability that assesses the frequency and gravity of ten neuropsychiatric domains including apathy, anxiety, delusions, hallucinations and aberrant motor activity. The study sample consisted of 25 healthy control subjects and 44 MS patients. The authors concluded that neuropsychiatric symptoms were present in 95 % of patients and 16 % of controls, and that the most severe symptoms were depression (79 % of patients vs. 16 % of controls), agitation (40 % vs 0 %), anxiety (37 % vs 4 %) and irritability (35 % vs 8 %). Hallucinations and delusions were present in 10 % and 7 % of patients respectively.

To identify variants of personality in terms of MS-associated maladaptation, Reznikova et al. [33] used a different, and interesting approach. The Standardized Multifactorial Personality Test (SPMT) [34], a modified version of the Minnesota Multiphasic Inventory (MMPI) [35], was administered to 34 patients, independently from the presence of florid psychopathology. The results showed a "psychotic variant" of maladaptation, specifically characterized by emotional impulsivity, excitatory features, hyperevaluation, over-responsiveness and lack of self-criticism.

14.4 Therapy

The review of the literature does not yield any specific guidelines, and the treatment of psychotic symptoms as an MS manifestation remains a therapeutic challenge.

There is no first choice drug. Published cases and case series suggest that the mainstay for treating acute exacerbations remain atypical antipsychotic agents at conventional dosages. The choice of a specific drug strictly depends on its therapeutic efficacy and on the adverse effects on the individual patient. The use of risperidone, olanzapine [36], quetiapine, clozapine [37], less frequently aripiprazole and ziprasidone, is reported [38].

In case of therapeutic failure of these medications, a diagnostic hypothesis of relapse may be considered and a high dosage of corticosteroids can be used [36, 39]. However, corticosteroid-induced psychosis directly related to dosage is widely described, and it is well recognized that corticosteroids can worsen the psychotic symptoms of MS. Dosage reduction or controlled withdrawal of the drug is usually an effective approach [16].

More recently, as a further therapeutic option, Gabelic et al. [16, 40] have proposed the use of plasma exchange. The authors described the case of a 25-year-old man with MS who, in course of corticosteroid treatment, developed a full-blown picture of paranoid psychosis with suicidal attempt unresponsive to therapy with quetiapine and fluvoxamine. The patient underwent treatment with five cycles of plasma exchange, and 6 days later was free from any psychiatric symptoms or signs.

Moreover, Lo Fermo et al. have suggested the use of mood stabilizers (e.g. carbamazepine, sodium valproate) in monotherapy or in association with antipsychotics [16].

Finally, psychotherapy can be proposed for some selected patients.

14.5 Neuroimaging

There is a paucity of information about the etiological association between the pathological process of MS and psychosis, and there is still no answer to the question of whether a single lesion can cause psychosis in MS patients.

Many case reports found a high lesion load in the periventricular frontal and temporal regions. In 1992, Feinstein et al. [41] found that MS patients with psychotic symptoms showed periventricular temporal lesions twice more frequently than MS patients without psychiatric manifestations. Diaz-Olivarrieta et al. [22] found that, if compared to normal controls, MS patients with hallucinations and delusions showed a significantly higher lesion load in fronto-temporal regions. More recently, Castellanos-Pinedo [39] described active lesions in the left hippocampus of one MS patient with delusions associated to somatic symptoms. Ron et al. [26] found no association between global measures of psychiatric disability and MRI changes, but reported a correlation between the degree of pathology in temporo-parietal regions and delusions. In their single patient, Gabelic et al. [40] also found a mesencephalic lesion, which may have contributed to psychosis onset through the interruption of the mesocortical limbic dopamine circuit. Finally, in four patients with acute exacerbation of psychotic symptoms, Blanc et al. [3] described active lesions in the frontal lobes and, in one patient with depressive symptoms followed by catatonia, also cerebellar lesions. Blanc et al. [3] also underlined that psychosis in course of MS could be explained by white matter damage that is more diffuse than what is recognized in MRI.

References

1. Cottrell SS, Wilson SAK (1926) Affective symptomatology of disseminated sclerosis; studies of 100 cases. J Neurol Psychopathol 7:1–30
2. Mitelman SA, Newmark RE, Torosjan Y, Chu KW, Brickman AM, Haznedar MM, Hazlett EA, Tang CY, Shihabuddin L, Buchsbaum MS (2006) White matter fractional anisotropy and outcome in schizophrenia. Schizophr Res 87(1–3):138–159
3. Blanc F, Berna F, Fleury M, Lita L, Ruppert E, Ferriby D, Vermersch P, Vidailhet P, de Seze J (2010) Evenements psychotiques inauguraux de Sclerose en plaques? Rev Neurol 166:39–48
4. Miller BJ, Buckley P, Seabolt W, Mellor A, Kirkpatrick B (2011) Meta-analysis of cytokine alterations in schizophrenia: clinical status and antipsychotic effects. Biol Psychiatry. 2011 Oct 1; 70(7):663–71. Epub 2011 Jun 8
5. Clerici M, Saresella M, Trabattoni D, Speciale L, Fossati S, Ruzzante S, Cavaretta R, Filippi M, Caputo D, Ferrante P (2001) Single-cell analysis of cytokine production shows different immune profiles in multiple sclerosis patients with active or quiescent disease. J Neuroimmunol 121:88–101
6. Hutchinson M, Stack J, Buckley P (1993) Bipolar affective disorder prior to the onset of multiple sclerosis. Acta Neurol Scand 88(6):388–393
7. Monaco F, Mutani R, Piredda S, Senini A (1980) Psychotic onset of multiple sclerosis. Ital J Neurol Sci 1(4):279–280
8. Reimer J, Aderhold V, Lambert M, Haasen C (2006) Manifestation of multiple sclerosis with paranoid hallucinatory psychosis. J Neurol 253:531–532
9. Fragoso D, Brooks B (2009) Psychiatric manifestations in multiple sclerosis patients and multiple sclerosis in psychiatric patient. Arq Neuropsiquiatr 67(4):1167–1168
10. Lyoo IK, Seol HY, Byun HS, Renshaw PF (1996) Unsuspected multiple sclerosis in patients with psychiatric disorders: a magnetic resonance imaging study. J Neuropsychiatry Clin Neurosci 8(1):54–59
11. Surridge D (1969) An investigation into some psychiatric aspects of multiple sclerosis. Br J Psychiatry 15:749–764
12. Skegg K, Corwin PA, Skegg DC (1988) How often is multiple sclerosis mistaken for a psychiatric disorder? Psychol Med 18(3):733–736
13. Stenager E, Jensen K (1988) Multiple Sclerosis: correlation of psychiatric admissions to onset of initial symptoms. Acta Neurol Scand 77:414–417
14. Patten S, Svenson L, Metz LM (2005) Psychotic disorders in MS: population-based evidence of an association. Neurology 65:1123–1125
15. ICD-9-CM (2002) "International Classification of Diseases, 9th revision, Clinical Modification"
16. Lo Fermo S, Barone R, Patti F, Laisa P, Cavallaro TL, Nicoletti A, Zappia M (2010) Outcome of psychiatric symptoms presenting at onset of multiple sclerosis: a retrospective study. Mult Scler 16(n6):742–748
17. Jongen P (2006) Psychiatric onset of multiple sclerosis. J Neurol Sci 245(1–2):59–62

18. Asghgar-Alì TK, Hurley R, Hayman A (2004) Pure neuropsichiatric presentation of multiple sclerosis. Am J Psychiatry 161:226–231

19. Kwentus JA, Hart RP, Calabrese V, Hekmati A (1986) Mania as a symptom of multiple sclerosis. Psychosomatics 27(10):729–731

20. Polman Ch, Thompson AJ, Murray TJ, McDonald WI (2001) Multiple sclerosis: the guide to treatment and management, 5th edn. Demos Medical Publishing, New York, pp 7–43

21. Matthews WB (1979) Multiple Sclerosis presenting with acute remitting psychiatric symptoms. J Neurol Neurosurg Psychiatry 42:859–863

22. Diaz-Olavarietta C, Cummings J, Velazquez J, de al Cadena C (1999) Neuropsychiatric manifestations of MS. J Neuropsychiatry Clin Neurosci 11(1):51–57

23. Smith EJ (2009) Multiple sclerosis presenting with erotomanic delusions in the context of 'don't ask, don't tell'. Mil Med 174(3):297–298

24. Sidoti V, Lorusso L (2007) Multiple sclerosis and Capgras' syndrome. Clin Neurol Neurosurg 109(9):786–787

25. Reimer J, Aderhold V, Lambert M, Haasen C (2006) Manifestation of multiple sclerosis with paranoid hallucinatory psychosis. J Neurol 253:531–532

26. Ron MA, Logsdail SJ (1989) Psychiatric morbidity in multiple sclerosis: a clinical and MRI study. Psychol Med 19(4):887–895

27. Shoja Shafti S, Nicknam Z, Fallah P, Zamani L (2009) Early psychiatric manifestation in a patient with primary progressive multiple sclerosis. Arch Iran Med 12(6):595–598

28. Agan K, Gunal DI, Afsar N, Tuncer N, Kuscu K (2009) Psychotic depression: a peculiar presentation for multiple sclerosis. Int J Neurosci 119(11):2124–2130

29. Hung YY, Huang TL (2007) Lorazepam and diazepam for relieving catatonic features in multiple sclerosis. Prog Neuropsychopharmacol Biol Psychiatry 31(7):1537–1538

30. Mendez MF (1999) Multiple sclerosis presenting as catatonia. Int J Psychiatry Med 29 (4):435–441

31. Monaco F, Mutani R, Piredda S, Senini A (1980) Psychotic onset of multiple sclerosis. Ital J Neurol Sci 1(4):279–280

32. Cummings JL, Mega M, Grey K (1994) The neuropsychiatric inventory: comprehensive assessment of psychopathology in dementia. Neurology 44:2308–2014

33. Reznikova TN, Terent'eva IY, Kataeva GV (2007) Variants of personality maladaptation in patients with multiple sclerosis. Neurosci Behav Physiol 37(8):747–754

34. Sobchik LN (2002) A standardized multifactorial method for studying personality (SMPT). Rech', St. Petersburg

35. Hathaway SR, Mc Kinley JC (1989) MMPI-2: Minnesota multiphasic personality inventory-2: manual for administration and scoring. University of Min&SHY; Nesota Press, Minneapolis

36. Thöne J, Kessler E (2008) Improvement of neuropsychiatric symptoms in multiple sclerosis subsequent to high-dose corticosteroid treatment. Prim Care Companion J Clin Psychiatry 10 (2):163–164

37. Chong SA, Ko SM (1997) Clozapine treatment of psychosis associated with multiple sclerosis. Can J Psychiatry 42(1):90–91

38. Davids E, Hartwig U, Gastpar M (2004) Antipsychotic treatment of psychosis associated with multiple sclerosis. Prog Neuropsychopharmacol Biol Psychiatry 28(4):743–744

39. Castellanos-Pinedo F, Galindo R, Adeva-Bartolomé MT, Zurdo M (2004) A relapse of multiple sclerosis manifesting as acute delirium. Neurologia 19(6):323–325

40. Gabelić T, Adamec I, Mrđen A, Radoš M, Brinar VV, Habek M (2011) Psychotic reaction as a manifestation of multiple sclerosis relapse treated with plasma exchange. Neurol Sci 33 (2):379–382

41. Feinstein A, du Boulay G, Ron MA (1992) Psychotic illness in multiple sclerosis. A clinical and magnetic resonance imaging study. Br J Psychiatry 161:680–685

Euphoria, Pathological Laughing and Crying

15

Silvia Romano and Ugo Nocentini

Many literature data have described both euphoria and pathological laughing and crying in MS patients.

The term euphoria is used to indicate a change in mood characterized by joy, exuberance and happiness connected to a successful and rewarding event, rarely associated with psychomotor excitement. Euphoria is considered a pathological condition when the emotional resonance is disproportionate to the real situation; it differs from mania as described in patients with bipolar disorders, because the mood change is stable and patients do not exhibit thought acceleration or the incessant impulse to think up new ideas and perform new activities. In MS patients, the description given by Cottrell and Wilson in 1926 [1] is still considered valid: they defined euphoria as a mental state characterized by expressions of cheerfulness, happiness and ease; patients appear serene and cheerful, and while aware of the disability that they suffer, report feeling healthy and fit, exhibiting an optimistic attitude towards the future and the prospects of healing.

There is no consensus on the prevalence of euphoria in MS: frequencies ranging from 0% to 63% are reported [1, 2] and this wide variation is the result of both differences among patients in terms of disease severity and duration and improper use of the term euphoria to describe all forms of emotional disorders. In fact, more recent data show a frequency of 10% [3]. The few clinical studies have reported that euphoria occurs more frequently in patients with high disability, long disease duration, chronic progressive MS and severe cognitive impairment [4]. On the other hand, neuroimaging studies show a correlation with ventricular enlargement and severe lesion load [3].

S. Romano (✉)
Centre for Experimental Neurological Therapies (CENTERS), Neurology Unit,
S. Andrea Hospital, Rome, Italy
e-mail: silvia.romano@uniroma1.it

U. Nocentini
Dipartimento di Neuroscienze, Università degli Studi di Roma "Tor Vergata", Rome, Italy
e-mail: u.nocentini@hsantalucia.it

U. Nocentini et al. (eds.), *Neuropsychiatric Dysfunction in Multiple Sclerosis*,
DOI 10.1007/978-88-470-2676-6_15, © Springer-Verlag Italia 2012

The interpretation of euphoria considered as a pathological condition has encountered significant changes over time. The first authors considered it to be a psychopathological disorder typical or pathognomonic of MS, but since the evaluation of cognitive functions has become even more systematic and specific, euphoria has been regarded as a consequence of cognitive impairment or one of the consequences of the loss of reasoning skills resulting from severe involvement of the frontal lobes and their connections [5]. It is surprising that there have been no recent studies exploring these aspects.

Pathological laughing and crying is a condition characterized by sudden, uncontrollable and incongruous episodes of laughing and crying that may occur alternately and are not associated to any stimulus. The patient may present a crisis of uncontrollable laughing or crying in the absence of a matching mood state and real motivation, or in the presence of stimuli that before the onset of the disorder would not have triggered this emotional reaction. Sometimes the stimulus may also have an emotional value opposite to the triggered emotional expression; for example, the patient may laugh in response to sad news or cry in response to a common action such as a hand movement [6].

Currently there is a tendency to consider pathological laughing and crying as being linked to an emotional disorder rather than a mood disorder, in which the episodes of laughing and crying, even though sudden or excessive, are always appropriate to the context. Although this disorder may overlap with a state of emotional lability, the terms are not synonymous because in the latter condition the episodes of laughing and crying are always triggered by appropriate stimuli. Care must be taken in differentiating this disorder from episodes of laughing and crying due to psychosis, substance use or personality disorders.

Pathological laughing and crying (also called pseudobulbar affect) have been described in several CNS diseases, such as MS [7], amyotrophic lateral sclerosis [8], gelastic epilepsy [9], Alzheimer's disease [10], stroke [11] and brain tumors [12]. In patients with cerebro-vascular lesions this disorder was associated with the involvement of cortico-bulbar tracts, especially at the level of the internal capsule, cerebral peduncles, and basis pontis, areas that control the movements necessary for laughing and crying [13].

Pathological laughing and crying is one of the four most common affective disorders described in MS, and constitute a symptom which is usually stressful and socially incapacitating for the patient.

Differences in the definition of the disorder, diagnostic criteria, and examined populations in different studies partially explain the great variability in the prevalence data reported in some studies (variable frequency from 7% to 95% in MS).

Cottrell and Wilson [1] were the first authors to be interested in this disorder; they observed a cohort of 100 MS patients and found that 71% constantly laughed and cried, 19% constantly had episodes of both crying and laughing, 3% constantly cried and 2% had a sudden shift from one condition to the other. Although the study analyzed the problem in detail, there are numerous methodological flaws that put the results into perspective: patients were enrolled at a tertiary referral center without specified selection criteria or disorder classification.

In 1941, Langworthy et al. [14] conducted a study in 199 outpatients with MS, reporting that in the advanced stages of the disease some patients showed pathological laughing and crying; the authors classified these symptoms as part of pseudo-bulbar palsy. This disorder was present in 13 patients with a prevalence of 6.5%, a much lower frequency than previously reported. Most of patients had episodes of uncontrollable crying, while some rapidly shifted from laughing to crying episodes.

Sugar and Nadel [15] analyzed 28 MS in patients with a long history of the disease and found that 79% had exaggerated emotional expressions; they also noted that 43% of patients constantly smiled or cried, 25% showed a mixed picture (smiling, laughing and crying), 7% constantly cried and 4% changed rapidly between the two states. However, the study has methodological flaws such as small sample size.

A subsequent study conducted by Pratt [16] compared 100 outpatients with MS (excluding more advanced forms of the disease) to a control group of 100 patients with other organic CNS diseases using the same interview administered by Cottrel and Wilson [1]. The study results showed that MS patients had affective disorders in 53% of cases, manifesting most commonly pathological laughing (22%) or crying (29%) compared to the control group, and that these disorders were directly correlated with the degree of disability and cognitive impairment.

Diagnostic criteria by Poeck [6], defining pathological laughing and crying as (a) sudden loss of emotional control on several occasions during one month; (b) episodes that occur in response to non-specific stimuli; and (c) absence of correlation with corresponding mood changes, have allowed a better assessment of the prevalence of this disorder by discriminating pathological laughing and crying from emotional lability.

Surridge [17] analyzed the frequency of psychiatric disorders in 108 MS patients and 39 controls with myotonic dystrophy, a neurological disease with high disability but without cerebral involvement. The study results showed that 10% of the patients with MS manifested exaggerated emotional responses, while no case was described in the control group, suggesting that these disorders are directly related to an organic lesion rather than disability.

More recent studies that have used Poeck's criteria associated with appropriate clinical scales have confirmed these data, showing lower prevalence rates than those initially described by Cottrell and Wilson [1].

Feinstein and colleagues [7] analyzed a sample of 152 patients with MS, applying the same diagnostic criteria associated with the PLACS scale [18], a specific scale for the classification of pathological laughing and crying that evaluates the intensity of the disorder, relationship with external events, degree of voluntary control, incongruent mood and level of disorder-induced stress. A control group composed of 13 patients with MS without affective disorders was also introduced. All patients underwent a battery of neuropsychological tests, and the presence of anxiety and depression was assessed. Feinstein's case-control study reported a frequency of pathological laughing and crying of 9.9%, confirming previously reported frequencies [14, 17] and demonstrating that sex and age do not affect pathological laughing and crying. As reported for euphoria, this disorder occurs more frequently in patients with a long duration of disease, a chronic progressive course and greater disability. No association was found between pathological laughing and crying and brainstem involvement or mood disorders.

Currently, the exact neurological basis of pathological laughing and crying is not fully understood; according to one of the first hypotheses [19] the disorder is determined by the loss of voluntary inhibition of a presumed center controlling laughing and crying located in the brainstem (disinhibition hypothesis), able to regulate breathing and facial movements associated with laughing and crying. In physiological conditions, laughing and crying would presumably be regulated by two different anatomical pathways, one connecting as yet unknown brain regions with the brainstem laughing and crying centers and involved in involuntary control of facial and breathing movements, and the other connecting the motor areas with the brainstem centers and involved in voluntary control of facial and breathing movements. However, this hypothesis is unsatisfactory because it leaves many questions unanswered: for example, it does not explain why a patient with pathological laughing and crying manifests episodes of laughing and crying in response to the same stimulus, manifests emotions not appropriate to the stimulus, or can voluntarily mimic laughter or crying.

Currently, based on retrospective clinical studies and individual cases reflecting clinical experience, it is believed that this disorder is caused by a dysfunction in cerebral pathways involving the cortex [20], brainstem [21] and cerebellum [22].

The description of patients with ischemic lesions in the anterior brain areas, in which frontal release signs are associated with pathological laughing and crying, has confirmed the close relation between this structure and sub-cortical structures that control mood and emotions, suggesting a possible involvement of the prefrontal cortex in the etiology of this disorder [23, 24]. According to the study by Feinstein and colleagues [20], patients with MS and pathological laughter and crying show a severer deficit of frontal functions than patients without MS, and this result suggests an involvement of the frontal areas of the cortex. In particular, because the patients were not significantly different from the controls on the Wisconsin Card Sorting Test (WCST), a test that assesses frontal deficits limited to the prefrontal cortex, a possible role of the orbito-frontal cortex has been suggested.

A recent MRI study performed in 14 MS patients with pathological laughing and crying and 14 MS patients without this disorder has revealed the presence of a higher lesion load than controls in brainstem, inferior parietal lobe, bilateral middle-inferior frontal lobe and right medial superior frontal lobe in patients with pathological laughing and crying. This finding confirms the importance of the frontal cortex in the pathogenesis of pathological laughing and crying and suggests the presence of a complex neural network that involves the brainstem and also the parietal cortex [21].

Different medications have been proposed for the treatment of pathological laughing and crying, including low doses of amitriptyline, SSRIs, levodopa, and amantadine [25]. The recent approval of Nuedexta (a combination of Dextrome-thorphan, 20 mg and Quinidine, 10 mg) in the USA for the treatment of pathological laughing and crying changes the choice treatment for one of the most disabling and embarrassing problems caused by MS [26, 27].

References

1. Cottrel SS, Wilson SAK (1926) The affective symptomatology of disseminated sclerosis. J Neurol Psycopathol 7:1–30
2. Minden SL, Schiffer RB (1990) Affective disorders in multiple sclerosis: review and recommendations for clinical research. Arch Neurol 47:98–104
3. Fishman I, Benedict RH, Bakshi R, Priore R, Weinstock-Guttman B (2004) Construct validity and frequency of euphoria sclerotica in multiple sclerosis. J Neuropsychiatry Clin Neurosci 16:350–356
4. Meyerson RA, Richard IH, Schiffer RB (1997) Mood disorders secondary to demyelinating and movement disorders. Semin Clin Neuropsychiatry 2:252–264
5. Rabins PV, Brooks BR, O'Donnell P et al (1986) Structural brain correlates of emotional disorder in multiple sclerosis. Brain 109:585–597
6. Poeck K (1969) Pathophysiology of emotional disorders associated with brain damage. In: Vinken PJ, Bruyn GW (eds) Handbook of clinical neurology, vol 3. North Holland Publishing Company, Amsterdam, pp 343–367
7. Feinstein A, Feinstein K, Gray T, O'Connor P (1997) Prevalence and neurobehavioral correlates of pathological laughing and crying in multiple sclerosis. Arch Neurol 54:1116–1121
8. McCullagh S, Moore M, Gawel M, Feinstein A (1999) Pathological laughing and crying in amyotrophic lateral sclerosis: an association with prefrontal cognitive dysfunction. J Neurol Sci 169:43–48
9. Arroyo S, Lesser RP, Gordon B et al (1993) Mirth, laughter and gelastic seizures. Brain 116:757–780
10. Starkstein SE, Migliorelli R, Tesón A et al (1995) Prevalence and clinical correlates of pathological affective display in Alzheimer's disease. J Neurol Neurosurg Psychiatry 59:55–60
11. Morris PL, Robinson RG, Raphael B (1993) Emotional lability after stroke. Aust N Z J Psychiatry 27:601–605
12. Monteil P, Cohadon F (1996) Pathological laughing as a symptom of a tentorial edge tumour. J Neurol Neurosurg Psychiatry 60:370
13. Kim JS, Choi-Kwon S (2000) Poststroke depression and emotional incontinence correlation with lesion location. Neurology 54:1805–1810
14. Langworthy OR, Kolb LC, Androp S (1941) Disturbances of behavior in patients with disseminated sclerosis. Am J Psychiatry 98:243–249
15. Sugar C, Nadell R (1943) Mental symptoms in multiple sclerosis. J Nerv Ment Dis 98:267–280
16. Pratt RC (1951) An investigation of the psychiatric aspects of disseminated sclerosis. J Neurol Neurosurg Psychiatry 14:326–336
17. Surridge D (1969) An investigation into some psychiatric aspects of multiple sclerosis. Br J Psychiatry 115:749–764
18. Robinson RG, Parikh RM, Lipsey JR, Starkstein SE, Price TR (1993) Pathological laughing and crying following stroke: validation of a measurement scale and a double-blind treatment study. Am J Psychiatry 150:286–293
19. Wilson SAK (1924) Some problems in neurology. II: Pathological laughing and crying. J Neurol Psychopathol 6:299–333
20. Feinstein A, O'Connor P, Gray T, Feinstein K (1999) Pathological laughing and crying in multiple sclerosis: a preliminary report suggesting a role for the prefrontal cortex. Mult Scler 5:69–73
21. Ghaffar O, Chamelian L, Feinstein A (2008) Neuroanatomy of pseudobulbar affect: a quantitative MRI study in multiple sclerosis. J Neurol 255:406–412
22. Parvizi J, Anderson SW, Martin CO, Damasio H, Damasio AR (2001) Pathological laughter and crying: a link to the cerebellum. Brain 124:1708–1719

23. Langworthy OR, Esser FH (1940) Syndrome of pseudobulbar palsy Anatomic and physiologic analysis. Arch Intern Med 65:106–121
24. Ross ED, Stewart RS (1987) Pathological display of affect in patients with depression and right frontal brain damage. An alternative mechanism. J Nerv Ment Dis 175:165–172
25. Feinstein A, Ghaffar O (2010) Disorders of mood and affect in multiple sclerosis, in: Kesserling J, Comi G, Thompson AJ (eds) Multiple Sclerosis: Recovery of function and neurorehabilitation. Cambridge University Press, Cambridge, pp 183–189
26. Pioro EP, Brooks BR, Cummings J et al (2010) Dextromethorphan plus ultra low-dose quinidine reduces pseudobulbar affect. Ann Neurol 68:693–702
27. Garnock-Jones KP (2011) Dextromethorphan/quinidine in pseudobulbar affect. CNS Drugs 25(5):435–445

Emotions and Multiple Sclerosis

16

Ugo Nocentini

Due to the increased attention being given to the topics of depression, anxiety and other psychiatric disorders in patients with Multiple Sclerosis (MS), as well as the greater knowledge now available in these areas, we should expect more studies on the topic of emotional processing in patients with MS, with particular reference to authoritative reviews of psychiatric co-morbidity in MS. The authors of these studies [1] not only report that some emotional states, such as the experience of anger, are particularly frequent in patients with MS [2–4], but underline the need for further research due to the importance of the topic and the lack of empirical data.

There is no doubt that relationships exist between emotional alterations and psychiatric disorders. These relationships tend to be seen in terms of symptoms (emotional alterations) within a syndrome (depressive disorder), but are actually much more complex and articulated. Connections between causes can also be interpreted in different ways: e.g., the depressive state might be the cause of anger attacks, or a certain way of processing emotional stimuli might lead to the development of depression.

One way of clarifying these issues is to look at research findings on patients with diseases of the nervous system. Thanks to the latest neuroimaging techniques, the location of lesions and the functioning of cerebral areas that underlie emotional processing can be identified with great precision. Therefore, besides clarifying the genesis of some disorders presented by patients with neurological diseases like MS, these studies might also provide useful data for understanding the origin of psycho-emotional disorders.

In recent years, some research has been carried out to investigate these issues.

The first study we will consider investigated the processing profile of one of the basic emotions, i.e., anger, in patients with MS. Nocentini et al. [5] evaluated a sample of 195 patients with clinically defined MS on a scale that measured various

U. Nocentini (✉)
Dipartimento di Neuroscienze, Università degli Studi di Roma "Tor Vergata", Rome, Italy
e-mail: u.nocentini@hsantalucia.it

U. Nocentini et al. (eds.), *Neuropsychiatric Dysfunction in Multiple Sclerosis*,
DOI 10.1007/978-88-470-2676-6_16, © Springer-Verlag Italia 2012

aspects of anger. Participants were selected from a sample of 300 patients based on the absence of significant cognitive disorders, which could have distorted the results of the psycho-emotional evaluations. To evaluate cognitive functions, a test battery was adopted that was able to explore the cognitive areas most frequently affected by the pathological processes of MS (attentive functions and speed of information processing; mnesic functions, executive functions; visuo-spatial functions). Patients were also evaluated using a depression scale (CMDI) [6], a scale to assess anxiety (STAY Y) [7], and the STAXI, State-Trait Anger Expression Inventory [8]. The last scale provides information on levels of trait anger (i.e. the predisposition to react with anger to certain stimuli), state anger (i.e. the level of anger currently being experienced), externally expressed anger (i.e. anger towards persons or objects), suppressed anger and controlled anger.

Results showed that levels of trait anger, state anger or externalized anger were not significantly different in the patients with MS and the general population. Surprisingly, however, the patients with MS presented high levels of suppressed anger and low levels of controlled anger. This profile is not consistent with that found in the normal processing of anger. In fact, high levels of controlled anger should correspond to high levels of suppressed anger. The MS patients' profile could be due to neuro-anatomical damage interfering with the cortico - subcortical circuits, which underlie emotional processing.

This hypothesis has received support from another recent study. Passamonti et al. [9] submitted 12 patients with MS (relapsing–remitting type, not cognitively compromised) and 12 healthy control subjects to a functional MRI examination while they were presented with stimuli (faces) with emotional content (expressions of anger, joy, sadness) or emotionally neutral stimuli. The authors found that although the patients with MS showed no significant differences in performing the task compared with the healthy controls, they showed activation of more extensive portions of the cerebral areas involved in the tasks and reduced functional connectivity between the amygdala and the prefrontal areas, i.e., structures considered crucial in the processing of negative emotions.

In 2009, Krause et al. [10] reported results of a study using functional MRI to evaluate the correlations between a task involving the recognition of emotional facial expressions and lesional and functional data. The authors studied three groups of subjects: MS patients with and without affective disorders and a group of healthy volunteers. The activation maps obtained during the presentation of faces with "unpleasant" emotional expressions (sadness, fear, anger) showed reduced activation of the ventral-lateral portion of the prefrontal cortex and the insula in the left hemisphere in patients with affective disorders. Reduced activation in these cortical areas correlated with the presence of lesions in the left temporal white matter, which could be at the base of the functional disconnection of the emotion circuits.

Jehna et al. [11] also found increased activation of specific cerebral areas in MS patients with no emotional, visual or cognitive dysfunctions after they had been shown faces with emotional expressions. In this case the emotions were anger, disgust and fear. For fear and disgust, excessive activation was observed in the posterior portion of the cingulate cortex and the precuneus.

Preliminary data gathered by researchers in the Neuroimaging laboratory of the IRCCS Santa Lucia Foundation confirmed the presence of higher levels of activation in other cerebral structures, i.e., in the medial frontal gyrus and the orbito-frontal cortex, respectively for anger and joy, and the anterior portion of the cingulate cortex and the mesial frontal gyrus for sadness.

The data reported in these studies on increased activation of cerebral areas after viewing emotional stimuli are consistent, but differences in the areas identified must be clarified in future investigations.

The greater and more extensive activation of cerebral areas could be the consequence of functional compensation to carry out specific activities in the presence of damage caused by MS. The spread of the response to emotional stimuli into areas not involved in healthy subjects could lead to psychopathology.

The last study we wish to discuss investigated the comprehension of emotional stimuli and the ability to use "theory of mind". The latter refers to the ability (probably unique to humans) to create hypotheses (theories) about thoughts present in the minds of others. This function has an important role in inter-personal and social interactions, is crucial for survival in articulated and complex communities, and seems to be linked to emotional processing. Henry et al. [12] submitted patients with MS and healthy controls to a series of tests. Some of the tests explored the ability to formulate a theory of mind and recognize emotions expressed by faces. The authors found that the patients with MS had deficits in recognizing emotions and in formulating a theory of mind. Performances on these trials correlated with those on trials that explored executive functions and speed of information processing. The correspondences between certain cognitive performances and the ability to formulate or use a theory of mind point to a common origin of the deficit in the functioning of the cortical-subcortical circuits that control both emotional processing and executive functions, i.e., circuits in which the amygdala and prefrontal areas predominate. Furthermore, as held by several researchers (e.g. [13]) the reciprocal role of emotions and reason in guiding behaviour can no longer be interpreted in terms of interference of the former with the latter or control of the latter by the former. Indeed, it has been demonstrated that deficits in emotional processing lead to a loss or a decrease of the ability to choose between profitable and damaging behaviours; and this difficulty could be due to the limits caused by deficient emotional processing on the construction of a theory of mind.

The data deriving from the studies we have discussed merely confirm the importance of the topic of emotional processing and correlated functions, and that it is not only necessary but also possible and fruitful to study these issues in patients with MS.

It is likely that the coming years will see the development of research in this field. One interesting way of investigating these topics more thoroughly could be to compare patients with different neurological pathologies, taking account of the differences in lesions that make these pathologies useful models of the different involvement of cerebral structures (e.g. by comparing models of cortical and subcortical pathologies).

References

1. Mohr DC, Cox D (2001) Multiple sclerosis: empirical literature for the clinical health psychologist. J Clin Psychol 57:479–499
2. Minden SL (1992) Psychotherapy for people with multiple sclerosis. Neuropsychiatry 4:198–213
3. Pollin I (1995) Medical crisis counselling: short-term therapy for long-term illness. Norton, New York
4. Mohr DC, Dick LP (1998) Multiple sclerosis. In: Camic PM, Knight S (eds) Clinical handbook of health psychology: a practical guide to effective interventions. Hogrefe & Huber, Seattle, pp 313–348
5. Nocentini U, Tedeschi G, Migliaccio R et al (2009) An exploration of anger phenomenology in multiple sclerosis. Eur J Neurol 16:1312–1317
6. Solari A, Motta A, Mendozzi L et al (2003) Italian version of the Chicago multiscale depression inventory: translation, adaptation and testing in people with multiple sclerosis. Neurol Sci 24:375–383
7. Spielberger CD, Gorsuch RL, Lushene RE (1989) STAI State-Trait Anxiety Inventory – Forma Y. Edizione italiana a cura di L Pedrabissi e M Santinello. Firenze, Organizzazioni Speciali
8. Spielberger CD (1992) STAXI State-Trait Anger Expression Inventory. Versione e adattamento italiano a cura di AL Comunian. Firenze, Organizzazioni Speciali
9. Passamonti L, Cerasa A, Liguori M et al (2009) Neurobiological mechanisms underlying emotional processing in relapsing-remitting multiple sclerosis. Brain 132:3380–3391
10. Krause M, Wendt J, Dressel A et al (2009) Prefrontal function associated with impaired emotion recognition in patients with multiple sclerosis. Behav Brain Res 205:280–285
11. Jehna M, Langkammer C, Wallner-Blazek M, Neuper C, Loitfelder M, Ropele S, Fuchs S, Khalil M, Pluta-Fuerst A, Fazekas S, Enzinger C (2011) Cognitively preserved MS patients demonstrate functional differences in processing neutral and emotional faces. Brain Imaging Behav 5(4):241–251
12. Henry JD, Phillips LH, Beatty WW et al (2009) Evidence for deficits in facial affect recognition and theory of mind in multiple sclerosis. J Int Neuropsychol Soc 15:277–285
13. Damasio A (1994) Descartes' error: emotion, reason, and the human brain. Putnam Publishing, New York

Part III

Cognitive Dysfunctions and Multiple Sclerosis

Cognitive Dysfunctions in Multiple Sclerosis

17

Ugo Nocentini, Silvia Romano, and Carlo Caltagirone

The presence of cognitive impairment in MS patients had already been identified by Charcot, as can be inferred from his descriptions of the disease dating back to 1877 [1]. After a long period in which these disorders did not receive appropriate attention from most authors, the last 30 years have seen remarkable progress in understanding their quantitative and qualitative characteristics. Data on the frequency of cognitive disorders are extremely variable and depend on the methodologies used and the type of patients investigated. According to the most methodologically correct studies, about 45–65% of MS patients show considerable cognitive dysfunction [2–6]. These dysfunctions range from selective disorders of specific functions to severe and widespread impairment. A recent study showed that the domains most frequently impaired in MS seem to be speed of information processing, and learning and memory, with a reported frequency of 51.9% and 54.3%, respectively. Overt dementia is rare in persons with MS [7].

Cognitive deficits are considered the main cause of the difficulties that patients encounter in their social and professional life [8]; in fact, most patients with cognitive impairment are unemployed and less involved in social and recreational activities. In addition, they are more dependent on others in their daily activities than MS patients without cognitive impairment [9].

Although the relationship between the clinical form of MS and severity of cognitive impairment is not yet completely clear, the most recent studies seem to confirm the hypothesis that there is a greater involvement in the CP than in RR forms [10–12].

U. Nocentini (✉) • C. Caltagirone
Dipartimento di Neuroscienze, Università degli Studi di Roma "Tor Vergata", Rome, Italy
e-mail: u.nocentini@hsantalucia.it; c.caltagirone@hsantalucia.it

S. Romano
Centre for Experimental Neurological Therapies (CENTERS), Neurology Unit, S. Andrea Hospital, University of Rome "La Sapienza", Rome, Italy
e-mail: Silvia.romano@uniroma1.it

U. Nocentini et al. (eds.), *Neuropsychiatric Dysfunction in Multiple Sclerosis*,
DOI 10.1007/978-88-470-2676-6_17, © Springer-Verlag Italia 2012

Regarding the correlations between the degree of physical disability (mainly measured using EDSS by Kurtzke [13]) and performance on neuropsychological tests, there are studies that have shown a positive correlation and studies that do not; the EDSS score in particular seems to have low predictive value in relation to cognitive performance [3]. In a recent study exploring patterns of cognitive impairment in 461 patients with RR MS [14], quite significant correlations were found between EDSS scores and performance on several neuropsychological tests; however, these correlations seem to be mainly the consequence of a large sample size and the statistical effects of a high number of correlations performed on the same sample. As regards this second aspect, the statistical corrections for repeated measures attenuated but probably did not delete the distortions; regarding the problem of sample size, the conclusion is that the predictive value of EDSS scores in relation to cognitive status is not useful for an individual patient. Currently no data are available on the predictive value, in relation to cognitive impairment, of other tools assessing general functional status in MS patients. The most recent attempt to overcome the problem consists in creating a tool that already contains a test for the evaluation of at least one aspect of cognitive functions: this is the Multiple Sclerosis Functional Composite (MSFC) [15], which includes the administration of a version of the PASAT, a test exploring attention and working memory. This test was chosen for its sensitivity and because it examines some functions commonly impaired in MS patients; however, there has already been a proposal to substitute the PASAT with another neuropsychological test that shows particular sensitivity and specificity in identifying cognitive impairment in MS patients, the Symbol Digit Modalities Test – SDMT [14, 16–18]. In fact, it has been argued that this test is the single most sensitive neuropsychological screener and has been shown recently to be the most sensitive test to detect cognitive deterioration over time [19]. In general, MSFC has proved to be the single best predictor of employment status in MS [20]. In fact, poor cognitive performance, particularly on measures of processing speed, verbal memory, and executive functioning, seem to significantly predict vocational status, even after controlling for the effects of age, education, gender, depression, disease course, and disease severity [21, 22].

With regard to correlations among the degree and characteristics of cognitive impairment and extent of anatomical lesions on MRI, there has been a considerable increase in studies dedicated to this subject in recent years. A specific and detailed discussion of the findings from these studies is beyond the scope of this book. We will therefore limit our observations to a brief excursus on successive developments in this field. Early studies showed no significant relationships between cognitive deficits and MRI data; an initial refinement of techniques and imaging processing methods demonstrated the existence of correlations among MRI parameters related to some brain regions and specific cognitive performance, in particular those involving frontal lobe functions. For example, significant correlations were found among the number of lesions (lesion load) localized in the left frontal lobe and the number of perseverative responses on the Wisconsin Card Sorting Test [23, 24]. Nevertheless, the results of other studies showed that these correlations could not be considered unique; in fact, the performance on tests exploring frontal functions showed correlations with both indices related to the extent of lesions in the frontal lobes and indices related to total lesion load and atrophy indices [25, 26]. As reported in the section on neuroimaging, over the years the availability of new techniques has made it possible to highlight the presence of pathologic changes, not only in the brain regions where the myelin component is prevalent: the involvement of the axonal component, and of cortical and deep nuclei gray matter such as the thalamus and basal ganglia, has been shown. Reports of correlations among the impairment of different cognitive functions and the degree of anatomical damage in these cortical and subcortical structures have consequently increased.

More recent years have seen the development of studies using functional MRI techniques to assess additional features of cognitive deficits and their relationship with patterns of activation of different brain structures.

For a brief but informative review on progress in the field of relationships between neuroimaging data and cognitive deficits, interested readers should refer to a recent review on the subject [27].

In any case, the recent wave of studies on the issue of cognitive impairment in MS patients confirms that some cognitive areas are more frequently affected (attention, memory, speed of information processing, executive functions, visual-spatial perception), and other skills (overall intellectual level and some components of memory and language) that are relatively well preserved.

17.1 Attention and Information Processing

Attention is one of the most frequently impaired cognitive functions in MS patients. It is a complex function with several components: attention processes involve the ability to direct and focus mental activity on set purposes, exercising control and integration functions on many other cognitive skills. A widespread and generally accepted clinical model divides attention into five subcomponents: focused attention, sustained attention, selective attention, alternating attention, divided attention. Focused attention means being able to respond to a signal in the absence of distractor stimuli. It is subdivided into: (a) tonic alert: the level of activation is always present; (b) phasic alert: increasing responsiveness in relation to an alarm stimulus or a signal. Sustained attention is the ability to maintain an adequate level of performance during a continuous and repetitive activity. In the case of disorders involving this attention component, there will be a progressive (time-on-task effect) and/or sudden (lapses of attention) deterioration in concentration. Selective or focal attention refers to the ability to isolate the target stimuli from distractor stimuli, while alternating attention refers to the ability to shift the attention focus from one task to another – an attention component that requires mental flexibility. Finally, divided attention makes it possible to simultaneously orient and focus mental activity on multiple stimuli. The concept of working-memory is closely linked to some of the attention components mentioned above. This function allows us to actively hold information in the mind for the time required to complete a particular task, and then direct attention to another task or return to a previous activity. The construct of working-memory includes a series of active control processes (e.g., strategies of repeating, encoding, organizing and retrieving information). It thus also depends on executive processes. Baddeley and Hitch [28] described a system related to the working-memory, called the central executive, which is the interface between the stock of long-term memory storage and working-memory. The attention components are all closely related to the working memory and executive processes.

Many studies have shown that information processing is compromised early in MS patients [14, 29–33]. Efficiency in information processing depends on the ability and speed of the brain structures to maintain and process information.

One of the first researches performed on these aspects of cognitive functioning was the study by Rao and colleagues [29], which found that MS patients required more time, compared to controls, to establish if a specific number was included or not in a series of numbers to remember; because the two groups had similar levels of accuracy, the authors suggested the presence of a deficit in information processing speed in the group of MS patients.

However, although many subsequent studies performed on this topic found important information about what seems to be one of the key problems of the disease and among the first symptoms to occur, a certain confusion has been generated between the terms "speed" and "accuracy of performance"; that is, it has not been possible to quantify the information processing speed by analyzing the accuracy of performance.

With regard to this experience, the objective of the research carried out by Demaree and colleagues [30] was to assess attention and information processing using appropriately modified tests so as to obtain a measure of the information processing speed and at the same time analyze the accuracy of performance. To perform this study, the authors used the PASAT test [34], considered particularly suitable for MS patients because it does not require the involvement of visual-motor skills. The commonly used version presents a variation of the administration time of numerical stimuli. Because in MS patients faster presentation rates are associated with decreased accuracy of performance, this test cannot measure the information processing speed and also analyse the accuracy of performance; the protocol purposely designed for this study was able to assess the accuracy levels of performance after establishing the optimal speed of stimulus presentation for each subject.

The results suggest that when there has been an adequate amount of time to encode the information, the accuracy of performance of MS patients is comparable to healthy controls. Therefore, the information processing speed would be a key factor affecting the working-memory encoding. Data from these studies suggest that, in addition to more impaired attention functions (divided and sustained attention), it is of fundamental importance in interpreting the results not to confuse "slow" performance with "poor quality" performance. Deficits in information processing speed seem to have a predictive value in relation to the progression of cognitive impairment. Deficits in information processing often occur together with deficits of different aspects of memory (working memory and long-term memory). It is not always easy to identify the direction of influence of one function on another: however, a higher frequency of deficits in information processing speed compared to deficits in working memory has been demonstrated [33].

With regard to the deficits of attentional components presented at the beginning of the paragraph, most data would exclude impairments of basal aspects of attention, such as the alert. On the other hand, deficits of other attentional components were also described in MS patients in the early stage. The study by Dujardin and colleagues [35] evaluated the ability of sustained and selective attention, simple and complex, in a group of patients with recent onset of MS using a program that included the search for target stimuli among some distractors on a screen. The patient was instructed to answer as quickly as possible and make the fewest errors. Response times and number of errors were calculated. The results of this study showed that patients with recent onset of MS had attentional disorders. Also in this case, as reported by Demaree and colleague [30], attentional disorders seem not so much to involve accuracy, which remains the same as in the healthy controls, as to be a direct consequence of the cognitive slowing identified in these patients. This slowing is, however, significant only in complex selective attention tasks that require a considerably higher cognitive load. However, it must be taken into account that the variability of the meaning attributed to the various terms used in the field of attentional functions by different researchers does not always allow the comparison or integration of results from different studies. Another aspect to consider in evaluating the efficiency of attention processes in MS patients is the influence of fatigue.

17.2 Memory

Memory is usually classified into short-term memory (STM) and long-term memory (LTM). The first has a limited capacity (the so-called short-term memory span) and allows information to be recorded for a limited period of time; LTM can store an unlimited amount of information for the entire life of an individual. Short-term memory should be distinguished from working memory [28, 36], which is the ability to keep present and active information from outside or LTM, for the time required to perform, step by step, specific complex tasks (e.g., to structure a speech, tackle and solve mental arithmetic, set up an activity). Working memory is a basic component of short-term memory; it analyzes information to be learned using an integrated and synchronized process, sorts them into logical sequences and thus facilitates learning. As already mentioned, the control of this functional component of memory, called the "Central Executive System" [28], has a role that goes far beyond the memory capacity in the strict sense, and is tightly integrated with attention, logic processing and planning capacity.

LTM is divided into explicit memory and implicit memory. The first allows learning, and conscious and aware retrieval of information. It is further divided into anterograde and retrograde episodic memory and semantic memory. Anterograde episodic memory allows the acquisition of new information, and is the most commonly damaged in amnesic patients; retrograde memory allows us to remember events that occurred in the past, and is usually partially preserved in patients with sequelae of head injury. However, it is always observed that the closer to the injury an event is, the greater the probability is that the subject does not remember it or remembers it only partially. Semantic memory, in contrast, concerns encyclopedic knowledge and information meaning, while implicit memory allows unconscious and involuntary learning. Examples of this type of memory are the priming phenomenon and procedural memory. A task that highlights the priming phenomenon is "stem completion": a person is asked to read a list of words and to give his opinion on the various words, then he is presented with the first three letters of a word and asked to complete the word; the subjects generally tend to complete the tasks using more frequently words belonging to the initial list than more commonly used words (if, for example, in the first list the patient reads the word LETTUCE, then he will tend to complete the initial letters LET with the word LETTUCE, rather than with the word LETTER). This type of memory is generally preserved even in cases of severe amnesic syndrome [37] and can be used to lead the patient to information storage or strategies of deficit compensation.

Memory is one of the most damaged cognitive abilities in MS patients and, also for this reason, it was and is the subject of numerous experimental and clinical studies. Not all memory components are, however, directly involved in the disease.

Working memory is usually affected as a direct consequence of generalized cognitive slowing [3, 29, 38, 39]. The aim of the study by Grigsby [39], performed on a group of 23 patients with chronic-progressive MS, was to evaluate the presence of working-memory deficits and to assess whether these deficits were largely due to the ability to process information; an information processing impairment would correlate with a working-memory deficit. The results showed that the scores obtained in the verbal fluency test (which also assesses capacity and processing speed) correlated significantly with all measures of short-term memory except the immediate recall of consonants. This result is consistent with the hypothesis that verbal information processing ability and speed are associated with performance on working-memory tests. Therefore, Grigsby and colleagues [39] have suggested that the main deficit observed in the disease consists in an impairment of information processing speed, but also of the capacity for central information processing. These data become clearly visible when patients are involved in complex tasks that require deeper information processing. The impairment of information processing is compatible with data indicating frontal lobe damage in MS.

Metamemory ability also appears to be impaired in patients with MS, because they tend to underestimate their memory disorders, thus manifesting unawareness of their memory efficiency [40]. Metamemory deficits correlate with the degree of executive function impairment.

Other researchers have revealed the presence of long-term episodic and semantic memory disorders in patients with MS. For example, according to Beatty and colleagues [41] and Jennekens-Schinkel and colleagues [42], the deficit is due to a difficulty in accessing information; in the list learning test, MS patients recall fewer items than controls, while the learning curves are similar in the two groups. MS patients with performances similar to those of controls on delayed recall tests, such as the short story test or word list test, do not show a higher rate of forgetting stored information, measured as percentage of forgotten information in delayed recall, compared with immediate recall. Rao and colleagues [43], using a test able to detect the different components of memory (encoding, storage, retrieval), have suggested that memory deficit is due to impairment of memory trace retrieval.

Subsequent studies [44, 45] seem to indicate, however, that in MS patients the acquisition deficit (encoding) of information prevails over recall (retrieval) deficit. The first study [44] aimed to clarify, through initial control of the amount of learned information, the encoding vs recall (acquisition vs retrieval) controversy. In this study, although the group of patients with MS required more presentations of the materials to learn the same amount of information, no significant differences were identified in the ability to retrieve or recognise the materials after an interval of 30 min compared with the control group.

Based on these results, the objectives of subsequent researches [45] have been: to replicate and explore in depth the results obtained in previous studies [44]; confirm the validity and applicability of the results, also regarding visual memory; assess the rate of forgetting information stored in long term memory. A modified version of a list learning test (Selective Reminding Test) with semantically related word pairs was used. In this version, the list of words is presented in full as long as the subject is not able to repeat all ten words of the test two consecutive times; this made it possible to control the acquisition difference between the group of MS patients and the control group. Efficiency and information processing speed were measured, using a variant of the PASAT attention test. The version used (AT-SAT) made it possible, as already mentioned with regard to the study by Demaree and colleagues [30], to control the presentation speed; for each patient the program sets the optimum inter-stimulus interval (threshold) that enables them to give a number of correct answers in at least 50% of cases. Visual learning was assessed using a modified version of the 7/24 Visual Memory Test, particularly suitable for the evaluation of patients with MS because not influenced by their degree of visual acuity or ability to motor control. Other neuropsychological tests were also administered to all study subjects.

The data from this study confirm that MS patients need more attempts to learn a list of words than controls do; this suggests that memory impairment is due to a deficit in the acquisition of information. In fact, after evaluating differences in the acquisition of verbal material, the group with MS did not differ from the control group in the number of retrieved words after 30, 90 min, or a week. Therefore, according to these studies, verbal memory impairment is not due to a deficit in recall from the long term storage, but in the initial acquisition of verbal material. In addition, there was no difference in the degree of verbal memory impairment between MS patients and control subjects; these data confirm that verbal information, once acquired, is recalled and recognized equally in the two groups, even a week after the first learning.

Regarding visual memory, the data emerging from the study [45] show a slightly different pattern: MS patients perform significantly worse than controls on visual memory tests, both in recall and in recognition at 30 and 90 min after acquisition. This result does not make it clear whether visual memory deficit depends on an impairment in storing, consolidating or recalling the memory trace. However, the rate of forgetting visual information previously acquired and stored seems to be similar between MS patients and controls.

Contrary to expectation, the information processing speed does not correlate with the number of attempts required to reach the criterion and most of the measures of recall and recognition in verbal memory tests. This does not agree either with the earlier study by De Luca and colleagues [44], which shows a correlation between performance on the PASAT and attempts to reach the criterion, or with other studies that suggest a relationship between processing speed and memory [38]. The PASAT is a test that requires speed, efficiency and flexibility of thought, among other functions. The modified version used for this study (AT-SAT) isolates the "speed" component of this difficult task. The results suggest, therefore, that processing speed alone does not explain memory and verbal learning performances; it can "contribute" to performing tasks requiring greater speed for an adequate performance and/or tasks requiring the simultaneous processing of multiple information (the dual-task paradigm). Therefore, the negative results of the study [45] agree with data obtained from a meta-analysis [46] in which a link was found between efficiency of information processing and memory performance in patients with MS. The association between performance on the PASAT and memory tests could be due to other cognitive components involved in the successful execution of the PASAT, such as flexibility of thought or ability to multitask.

Why, then, in patients with MS allowed multiple attempts to reach the criterion, do recall and recognition improve to the extent that they obtain results similar to those of the controls? According to DeLuca and colleagues [45], it is not the repeating trials that are responsible for performance improvement, but the better quality of information encoding derived from them. Encoding is defined in cognitive psychology as "the process that enables us to interpret and organize items in units of memory". It follows that the best organization of encoded material, as indicated by repeated learning opportunities, improves memory performance.

Data from this study [45] suggest that verbal memory and visual memory follow a different pattern of impairment. The hypothesis is also supported by data from a previous study by Rao and colleagues [2].

Like most amnesic subjects, MS patients too seem to retain implicit memory. Seinelä and colleagues [47] evaluated implicit memory ability in a group of MS patients with cognitive impairment. Implicit memory is usually measured using priming tasks; the level of accuracy or speed with which a memory task is performed may be facilitated by prior exposure to the type of information that will be required to carry out the task. One of the most widely used priming tests is "stem completion" (see above). Although explicit memory is severely impaired in amnesic patients, they show normal performance on implicit memory tasks such as "stem completion". The data emerging from this study [47] show that there is a dissociation between implicit memory and explicit memory in patients with MS who have cognitive impairment; they also seem to confirm the results of previous studies suggesting the presence of separate systems for explicit memory and implicit memory. Even in patients manifesting widespread cognitive impairment, the neuronal circuits involved in implicit memory do not seem to deteriorate.

17.3 Executive Functioning

The term "executive function" refers to a set of complex aspects of cognitive functioning: management initiative; ability to inhibit response; persistence on the task; planning; analysis and problem solving; abstract and conceptual thinking skills; management of cognitive resources. The correct functioning of these aspects requires that functions such as attention and memory are not impaired.

In practical terms, assessment of executive functions raises difficulties due to the dependence, mentioned above, of the efficiency of these functions on other levels of cognitive functioning, and the unclear definition and delimitation of the same executive functions.

However, a series of neuropsychological tests proposed for the assessment of executive functions [48, 49] would simultaneously examine multiple aspects of this functional area and provide an idea, sometimes general, of the efficiency of these functions. What is more frequently reported by people close to the patient than by the patients themselves corresponds to a performance deficit on these tests: difficulty in analyzing and solving problems that require the subjects to consider alternative solutions, difficulty in following a program of structured or new activities, tendency to persevere in an action even if it is obviously ineffective.

There is a general tendency to consider a deficit of executive functions as indicative of damage to the structures of the frontal lobes or connections between deep brain structures (e.g., basal ganglia) and the frontal lobes.

Thus, from the perspective of the anatomical location of the lesion, patients with MS present a significant risk of executive dysfunction.

In fact, Pearson and colleagues [50] had already shown significant differences between MS patients and healthy controls in a task requiring the identification of rules for problem solving; further studies [51, 52] have shown a dysfunction in the identification of elements matched to a series of objects and situations or relationships that link actions or statements according to a logical sequence (identification of concepts, ability to abstraction).

According to the results of some researches [2, 14, 16, 25, 40, 52, 53], the most evident dysfunction in MS patients seems to be the perseveration of ideas or solutions that are no longer appropriate to changing circumstances. Nevertheless, the results of studies that have examined separately the different aspects of executive ability seem to suggest that the greatest difficulty of these patients consists in a failure to identify concepts.

In fact, a critical examination of the various studies suggests some general considerations. Progress in defining the different subcomponents of executive functions and the different levels into which the problem-solving process can be divided (e.g., planning aimed at achieving a goal, generation of flexible strategies, maintenance of the set, monitoring of the action and inhibition of irrelevant stimuli), described in theory, have only recently been applied to detecting deficits of patients with MS: e.g., Birnboim and Miller [54] evaluated the ability to implement and maintain a work strategy for the effective execution of a task.

The results of this study are consistent with the possibility that MS patients have difficulty in facing new situations, completing a project, dealing with complex information. It is to be noted that Birnboim and Miller [54] described no significant differences between patients with RR and SP MS form in terms of ability to apply strategies.

These data have been confirmed by a recent study on the population of New Zealand. Drew and colleagues [55], analyzing a sample of 95 patients with different forms of MS (RR, SP, chronic progressive and benign) reported that 17% of patients showed difficulties in executive functioning (such as for example the generation of flexible strategies, inhibition and fluency) of variable severity but without significant differences between the different forms. The published data suggest that, to evaluate executive functions, solutions are required that can separate the aspects actually related to executive functions from the components related to different levels of cognitive functioning. The impairment of various aspects of executive functions is not uniform in patients with MS, and the pattern of these deficits does not correspond to what is found in patients with frontal damage due to other etiology. In fact, the deficits in executive functioning described in MS patients are not exclusively due to damage in the frontal lobes: some authors [23, 24], on the basis of correlation studies, have hypothesized a close relationship between frontal lobe damage identified by MRI and deficits of these functions; other studies [25, 56] have again described the difficulty of establishing the specific contribution of frontal lobe pathology in determining executive deficits in the presence of widespread brain damage such as occurs in MS.

It is clear that further research is needed to define more precisely the characteristics of executive deficits in patients with MS and to understand the main causes of these deficits. Nevertheless, at the clinical practice level, while taking account of the limits mentioned, this aspect of cognitive functioning must also be evaluated in the individual patient with MS. Finally, it is important to attempt to limit the consequences of any executive deficits, considering the repercussions that they may have on the patient's daily life, including, not least, those relating to treatment decisions and the planning of intervention strategies.

17.4   Visuo-spatial Functions

The assessment of visuo-spatial functions in patients with MS has been the subject of a very limited number of specific researches. This corresponds to the difficulties that, even at the clinical level, are encountered in assessing aspects connected with the visual sensory impairment that is so frequent, and sometimes severe, in patients with MS.

Despite these difficulties, Vleugels and colleagues [57, 58] reported the results of an extensive research on visual-perceptual impairment in patients with MS which has become an important reference point.

Vleugels and colleagues [57, 58] administered 31 neuropsychological tests able to evaluate both spatial and non-spatial visual-perceptual skills to a sample of 49 MS patients with no significant visual impairments. They found a quite significant (26%) percentage of patients with impaired performance in 4 or more tests, but there was no uniformity or selectivity in the type of impairment. Significant levels of impairment were found only in 4 of the 31 tests: a color discrimination test, a test for the perception of visual illusions, a test for the perception of objects, and a test assessing the associative stage of the visual perception of objects. Confirming the variability of visual-spatial deficits in patients with MS, although the 4 tests showed a good predictive power of the overall visual-spatial deficits as determined on the basis of all 31 tests, they were not completely satisfactory as regards sensitivity and specificity towards an independent criterion for assessing visual-perceptual impairment.

Performance on visual-perceptual tests correlated weakly with general cognitive status, global physical disability measured by EDSS [13], and EDSS scores for pyramidal, cerebellar and brainstem functional systems; no significant correlations were found with other neurologic signs, disease duration, type of MS course, history of optic neuritis, depression or drugs.

On the basis of the results of this study, it cannot be ruled out that the deficits identified in visual-perceptual tests were due to visual deficits not detectable using the usual procedures, or clinical deficits of other cognitive functions. The same authors accordingly performed an additional study [58] subjecting patients with MS to both an extensive battery of visual-perceptual tests and tests designed to assess spatial and temporal resolution skills in relation to visual stimuli and evoked potentials elicited by changing visual patterns. The results of this detailed research support the hypothesis of the substantial independence of visual-perceptual deficits compared with other cognitive and visual deficits. The weak correlation between performance on visual-perceptual tests and the time resolution for visual stimuli (sustained, probably, by the magnocellular system and dorsal visual projections) requires further clarifications of the causal mechanisms of visual-spatial deficits, excluding the primary visual deficits that very frequently occur in MS patients.

17.5 Language

There are very few studies that have explored, directly and in detail, the language skills of MS patients [59, 60].

There do exist some sporadic reports on acute aphasic syndromes resulting from wide demyelinating lesions in the white matter below the cortical areas assigned to language elaboration processes, or from lesions that disconnect these areas from each other or from the areas of visual and auditory perception [61–64]. Because they are single reports, they cannot be considered decisive in determining whether, in MS, language disorders are a significant problem.

The systematic studies [59, 60] have shown deficits in oral naming, verbal fluency (phonetic and categorical), auditory comprehension (in particular of complex or ambiguous material in terms of syntax and grammar [65]), and reading. Henry and Beatty confirmed these results, demonstrating that patients with MS had impaired phonemic and semantic fluency, and that these deficits correlated with the degree of neurological impairment [66].

Deficits in linguistic tasks may be a consequence of a primary impairment. However, they might also be caused by deficits of other aspects of cognitive function (attention and information processing, memory, executive functions) whose efficiency is certainly also relevant to language. Other deficits could be due to an altered information transmission between the two hemispheres, resulting from frequent lesions of the corpus callosum; this transmission is probably needed for more complex linguistic processes. The studies conducted so far are inconclusive regarding the hypotheses on the interpretation of language deficits in patients with MS. Theoretical elements suggest that disconnection disorders have a higher frequency than has been observed so far. It is not easy to identify any language deficits in these patients on routine clinical examination: they tend to be mild or moderate and are related to more complex levels of linguistic elaboration. On the other hand, in the formal neuropsychological evaluation of language functions, there is the question of the type of testing material to use, considering the sensory-motor problems of MS patients (e.g., visual acuity and discrimination).

The issue of language skills in patients with MS therefore remains open and should not be overlooked: the role of verbal mediation in the activities of daily living is by no means irrelevant.

17.6 General Intelligence

Manuals and articles describing specific research on the cognitive functions of MS patients state that general intelligence is sufficiently preserved in these patients. However, there are insufficient data to accept or reject this claim. There are considerable difficulties in establishing a shared definition of general intelligence and, therefore, in identifying appropriate measuring tools, and these difficulties probably discourage researchers from carrying out studies on the topic of intelligence.

With regard to MS patients, there are also difficulties caused by the presence of deficits of other skills that can affect performance on the tests routinely used for the evaluation of intelligence (e.g., The WAIS – Wechsler Adult Intelligence Scale – Revised [67]).

Consequently, the results of studies showing a decrease of IQ in patients with MS compared to control subjects [56] and to the so-called pre-morbid IQ [68] may have suffered from the influence of other cognitive impairments, the control of which is not at all easy when assessing intelligence.

The assessment of pre-morbid intelligence refers to the level of knowledge and skills acquired before the onset of a CNS disease. In the case of MS it is not easy to determine when the disease process began; it might have been many years before the onset of evident clinical manifestations.

So, in the case of MS patients, the theme of general intelligence has not yet found appropriate responses. A different theoretical approach to the concept of intelligence and the consequent creation of assessment tools will probably be necessary.

17.7 Neuropsychological Assessment in Multiple Sclerosis

The cognitive impairment that characterizes patients with MS can lead to significant disability, sometimes without significant deficits in other neurological functions. Therefore, it may be appropriate to assess cognitive functioning at the moment of diagnosis or even during the first episode suggestive of the disease.

The tools most frequently used in the evaluation of MS patients do not provide a formal evaluation of the patient's cognitive abilities, because this would require time and expertise which are rarely available. A precise identification of the cognitive patterns of each individual patient with MS must be made through a certain number of neuropsychological tests exploring different aspects of intellectual functioning. In practice, extensive neuropsychological assessments are too long and require too many resources to be routinely applied. It is thus necessary to have a brief screening battery, with a high specificity and sensitivity, which is fast and easy to administer and evaluate, and which can also be used by unskilled personnel and administered to patients with sensory-motor disabilities. The use of an extended battery of heterogeneous tests that can make an in-depth exploration of all the different mental skills contrasts with the choice of a small battery of shorter and more essential tests that can provide an overview of the patient's cognitive abilities. In this regard, the use of a screening battery (MMSE; SEFCI; Rao's Battery, etc.) allows an initial assessment of a greater number of patients, referring only those patients showing significant deficits to a more detailed evaluation process.

Based on these considerations, different evaluation tools have been proposed. In the first instance an attempt was made to apply a rapid assessment tool to the examination of MS patients, the Mini Mental State Examination (MMSE) [69], already widely used in screening for presenile and senile dementia. The test consists of five assays examining spatial and temporal orientation, attention, short-term and long-term memory, constructive-organizational skills and language, and takes about 5–10 min to perform. However, there have been many criticisms of the use of the MMSE for screening MS patients. The study by Beatty and Goodkin [70] found, for example, a very low degree of sensitivity for the presence of cognitive deficits in patients with MS. For this reason, researchers have thought it necessary to increase from 24 to 28 the score below which it is possible to hypothesize the presence of cognitive impairment. In addition, almost all MMSE validation studies were conducted on patients who presented cortical dementia features according to a common clinical definition. From a classical point of view, most patients with MS have a pattern of sub-cortical cognitive impairment. This pattern is characterized by an overall slowing of cognitive processes, memory disorders, difficulty in problem solving and affectivity disorders (apathy and depression), with substantial preservation of language, praxic and gnosic functions. These cognitive aspects are not assessed or are assessed only superficially by the MMSE. Subsequent studies have therefore been directed at developing new assessment tools that are sufficiently short and sensitive to cognitive impairment in MS.

Of these tools, the Screening Examination for Cognitive Impairment (SEFCI) is one of the most prominent [71]. This battery, developed specifically for MS patients, is one of the most widely used screening tools and has greater validity. It takes about 25 min for administration, even by appropriately trained unskilled personnel. This short battery consists of a series of tests investigating different cognitive functions: immediate and delayed memory (Short Word List), ability to name and verbal fluency (Shipley Institute of Living Scale), visual-spatial attention (Digit Symbol Modalities Test).

Another commonly-used tool, developed specifically for MS patients, is the Brief Repeatable Battery of Neuropsychological Tests (BRBNT) [72]. The battery consists of 5 tests investigating the following cognitive domains: short and long term memory capacity (Selective Reminding Test), short and long term visual-spatial memory capacity (10/36 Spatial Recall Test), ability to sustain attention and concentration (Paced Auditory Serial Addition Task), information processing speed and working memory (Symbol Digit Modalities Test), verbal fluency skills (Word List Generation). A fairly recent Italian study [73] showed that the BRBNT is slightly, but not significantly, more sensitive than SEFCI in identifying patients with cognitive impairments. The tests that best discriminate between MS patients and controls are: the Selective Reminding Test and the PASAT regarding the BRBNT and the SDMT for the SEFCI. However, the BRBNT takes about 11 min more for administration.

Another battery for cognitive performance screening of MS patients is the Neuropsychological Screening Battery for Multiple Sclerosis (NPSBMS) [2]. This battery includes a verbal learning test and a short and long term memory test (Selective Reminding Test), a spatial learning test (7/24 Spatial Learning Test), an attention and working memory test (Paced Auditory Serial Addition Test) and a verbal fluency test (Controlled Oral Word Association Test). The administration of this battery takes a short time (about 20 min) and does not require skilled personnel; it has demonstrated a sensitivity of 71% and a specificity of 94% in discriminating cognitively impaired from cognitively intact patients [2].

Another example of a battery applied to the evaluation of MS patients is the Repeatable Battery for the Assessment of Neuropsychological Status (RBANS) [74]. It assesses short and long term memory, language, attention and visual-spatial skills. It takes about 30 min for administration. This battery has been used in the evaluation of patients with other diseases involving cognitive deficits. It has excellent normative data for individuals from 20 to 89 years and, by applying a simple algorithm, makes it possible to evaluate the cognitive state of patients with Alzheimer's disease, Huntington's disease, subcortical vascular dementia and Parkinson's disease with dementia.

The results of the study by Aupperle and colleagues [75] showed that both the SEFCI and the NPSBMS are more likely to identify MS patients with cognitive impairment than the RBANS. Because it requires less time for administration, the SEFCI is preferable if the screening goal is to test a large number of patients; it has the limitation that its reliability is relative to a single assessment, for the reason that the effects of learning on repeated administrations are unknown. Both the NPSBMS and the BRBNT seem to adapt better to clinical studies that require multiple assessments over time. At our institute a battery composed of the Mental

Deterioration Battery [76] and two other tests, the Modified Card Sorting Test [77] and the Symbol Digit Modalities Test – Oral Version [78, 79], is used. This battery has also been used in a multicenter study involving more than 600 patients, of whom 461 with the RR form [14]. It contains tests on information processing speed, MBT, MLT, executive functions, visual-perception, language, and intelligence.

The data reported in the literature about the frequency of cognitive impairment in MS patients confirm the need for agile, reliable screening tools. Although there are a large number of neuropsychological tests to assess cognitive functions in MS, it is not easy to compose a suitable battery. In fact, the tests in this battery must allow an accurate investigation of the cognitive areas that are potentially compromised in MS patients, limiting as much as possible the concomitant sensory-motor deficits. In 2002 a panel of experts on the neuropsychological assessment of MS patients suggested minimum criteria for the composition of a battery of tests suitable for clinical use and therapeutic trials (MACFIMS) [80].

Very recently, a panel of experts in the research and assessment of MS cognition recommended the SDMT for a 5 min screening and the administration of the CVLT and the Brief Visuospatial Memory Test if the testing time could be expanded to 15 min. This suggested strategy will be tested through validation studies conducted in various languages [81].

If a cognitive impairment has been identified using a screening test, further evaluations of the cognitive functioning areas in which deficits have been highlighted must be performed. This detailed elaboration will be useful to understand the reasons for the difficulties that the patient encounters in the activities of daily life and to make the patient himself, his family and caregivers aware of these issues. The most detailed definition of the cognitive dysfunction is essential in planning and implementing a rehabilitation program for cognitive deficits, and to verify the results.

For a detailed assessment of one or another aspect of cognitive functioning several neuropsychological tests are available, a comprehensive description of which is outwith the scope of this chapter. For further information the reader should refer to specialized texts [48, 49].

At the end of this chapter on cognitive dysfunctions in MS and their evaluation, we want to return briefly to the issue of cortical and sub-cortical dementia. This distinction also involves the case of cognitive impairment in MS patients. The characteristic pattern of cortical dementia is typical of Alzheimer's disease, while that of sub-cortical dementia corresponds to impairment described in Huntington's disease [82]. The cognitive pattern of MS has long been considered similar to the sub-cortical dementia profile. This classification is consistent with the assumed localization of MS lesions in sub-cortical white matter areas. The progressive identification, by MRI and histopathology, of cortical and sub-cortical gray matter lesions resulted in a change of perspective. The neuropsychological data have confirmed to some extent that, in the case of MS, we are probably dealing with an even more complex cognitive impairment profile, in which many factors must be taken into account. A few of these are: the presence of cortical and sub-cortical lesions; phenomena resulting from multiple disconnections, which vary from one individual to another, among the cortical areas and among these and the sub-cortical structures; phenomena of plasticity and functional compensation; a different level of cognitive reserve. This last factor, referring to the concept that higher lifetime intellectual enrichment results in increased cerebral efficiency, has been evaluated in some recent studies by Sumowski and colleagues [83–85]. In these studies, MS patients with low cognitive reserve displayed significant cognitive impairment, while those with high cognitive reserve manifested little or no cognitive impairment, even with the same degree of brain atrophy. Higher intellectual enrichment seems to be able to lessen the negative impact of brain atrophy on both learning and memory, suggesting a protective role of cognitive reserve against disease-related cognitive impairment in MS patients.

In the next few years, research will probably confirm that the only way to have a reliable idea of cognitive impairment profiles in MS patients is to take into account all the presumed factors rather than a single point of view. We therefore prefer to not assign a particular pre-established label to cognitive deficits in MS patients, also because this attribution is not helpful in directing us toward possible solutions of the consequences of cognitive dysfunction.

17.8   The Relationships of Psychopathological Dysfunctions to Cognition

Interest in the relationships between psychopathological disorders and cognitive dysfunction arises from several considerations. The first is epidemiological: as reported in previous chapters, MS patients have a high prevalence of cognitive and mood disorders, two clinical features that cannot be overlooked. The second consideration relates to the possible interactions between mood disorders and cognitive impairment. If we consider the data on mood disorders in patients with no neurological disease, we see that these disorders can significantly affect cognitive performance, but it is also true that the perception of cognitive disorders can lead to psychopathological reactions. The third consideration is more connected to the following research topics: cognitive impairment in MS affects some functions more frequently than others; the impaired functions mainly involve frontal and temporal lobe structures; these areas play a decisive role in emotional elaboration; so the relationship between cognitive and emotional disorders may be related to a common cause of damage.

With regard to the above considerations, there is not much to add about the epidemiological data.

As for the second consideration, the qualitative and quantitative evaluation of the relationship between depression and cognitive functioning in patients with MS is a difficult task. As we said, it is not easy to determine the directionality of the influence of one type of disorder on another. The awareness, but also the (possibly wrong) assumption, of a cognitive deficit may certainly cause or worsen a mood alteration, but on the other hand the influence of depression on the cognitive abilities of the affected individual is known. The result of mood disorder evaluation may be influenced by many factors: in the very early stages of the disease the weight of some of these factors may be absent or poor (e.g., pharmacological effects; personal experience of the disease), but in these stages cognitive disorders are usually mild and not detected by the neuropsychological tests commonly used in clinical practice.

While we must take into account the methodological limitations of different studies, the most consistent results of research carried out on this issue in patients with MS seem to be those related to the negative influence of depression on the working memory and information processing, which are more demanding in terms of resources involved [46, 86, 87].

Regarding working memory, Arnett and colleagues [87] suggested that the aspect of this function most influenced by depression was the central executive component. However, the involvement of some aspects of executive functions as well as processing speed cannot be excluded in the relationships among attention, working memory and depression. The possible role of executive functions was subsequently evaluated in another study by the same group [88]: however, it has been shown that although the ability to plan explains part of the variance in depression levels, the function most closely related to depression is information processing speed.

Another study [89] showed that the influence of depression on information processing speed is maintained even after checking the accuracy parameter; the study confirms that to distinguish the influence of depression and fatigue on information processing speed is not easy. Depression and fatigue exert their greatest influence on the relationship between information processing speed and particularly demanding tasks (immediate recall of verbal information and word list learning).

As regards the other aspects of cognitive functioning, the data available do not allow us to draw meaningful conclusions, although there is a prevalent absence of significant influences [86].

If we go into details regarding the third consideration, the aspects of cognitive functioning most related to the presence of depression (working memory, attentional functions) are supported by frontal lobe structures, which also seem to be involved in the regulation of behavioral adjustment and emotional responses. The combination of depressive disorder and attention and working memory deficits might be due to the involvement, often significant in MS patients, of the frontal lobes and their connections with other brain structures.

This hypothesis is beginning to receive some confirmation from experimental data: Figved and colleagues [90] found an association between memory and processing speed deficits and depression, and a relationship between apathy and the intrusions in a word list recall task.

With regard to other aspects of psychopathology, data on possible relations with cognitive dysfunctions appear to be consistent. This might also be due to a more general lack of systematic assessment of the quantitative and qualitative features of disorders such as anxiety or psychosis in patients with MS. It has also been hypothesized that the presence of an anxiety state is considered "normal" by doctors but also by the patients themselves, to the point that it is not accurately investigated or reported with particular emphasis.

Finally, we should mention a condition that has interested researchers since the earliest systematic observations on the disease: euphoria. Data on the characteristics of this condition have been presented previously. Here we want to restate that the interpretation of euphoria has changed significantly over time. The first authors regarded it as a psychopathological disorder characteristic or pathognomonic of MS. As the assessment of cognitive functions has become more systematic and specific, euphoria has come to be considered rather as a consequence of cognitive impairment, or is placed among the consequences of the loss of critical skills due to severe involvement of the frontal lobes and their connections.

However, despite the presumed interest in exploring further the relationship between cognitive impairment and the state of euphoria, there are surprisingly no recent studies that have explored the issue.

References

1. Charcot JM (1877) Lectures on the diseases of the nervous system delivered at the Salpêtrière. The New Sydenham Society, London
2. Rao SM, Leo GJ, Bernardin L, Unverzagt F (1991) Cognitive dysfunction in multiple sclerosis. I. Frequency, patterns, and prediction. Neurology 41:685–691
3. Rao SM (1995) Neuropsychology of multiple sclerosis. Curr Opin Neurol 8:216–220
4. Fischer JS (2001) Cognitive impairment in multiple sclerosis. In: Cook SD (ed) Handbook of multiple sclerosis. Marcel Dekker, New York
5. Bobholz JA, Rao SM (2003) Cognitive dysfunction in multiple sclerosis: a review of recent developments. Curr Opin Neurol 16:283–288
6. Amato MP, Zipoli V, Portaccio E (2006) Multiple sclerosis-related cognitive changes: a review of cross-sectional and longitudinal studies. J Neurol Sci 25:41–46
7. Benedict RH, Zivadinov R (2011) Risk factors for management of cognitive dysfunction in multiple sclerosis. Nature Review Neurology 7:333–342
8. Rao SM, Leo GJ, Ellington L, Nauertz T, Bernardin L, Unverzagt F (1991) Cognitive dysfunction in multiple sclerosis. II. Impact on employment and social functioning. Neurology 41:692–696
9. Kesserling J, Klement U (2001) Cognitive and affective disturbances in multiple sclerosis. J Neurol 248:180–183
10. Amato MP, Ponziani G, Siracusa G, Sorbi S (2001) Cognitive dysfunction in early-onset multiple sclerosis: a reappraisal after 10 years. Arch Neurol 58:1602–1606
11. Huijbregts SCJ, Kalkers NF, de Sonneville LMJ, de Groot V, Reuling IEW, Polman CH (2004) Differences in cognitive impairment of relapsing remitting, secondary, and primary progressive MS. Neurology 63:335–339
12. Wachowius U, Talley M, Silver N, Heinze HJ, Sailer M (2005) Cognitive impairment in primary and secondary progressive multiple sclerosis. J Clin Exp Neuropsychol 27:65–77
13. Kurtzke JF (1983) Rating neurological impairment in multiple sclerosis: an expanded disability status scale (EDSS). Neurology 33:1444–1452
14. Nocentini U, Pasqualetti P, Bonavita S et al (2006) Cognitive dysfunction in patients with relapsing-remitting multiple sclerosis. Mult Scler 12:77–87
15. Polman CH, Rudick RA (2010) The multiple sclerosis functional composite: a clinically meaningful measure of disability. Neurology 74(Suppl 3):S8–S15
16. Parmenter BA, Weinstock-Guttman B, Garg N, Munschauer F, Benedict RH (2007) Screening for cognitive impairment in multiple sclerosis using the symbol digit modalities test. Mult Scler 13:52–57

17. Benedict RH, Duquin JA, Jurgensen S et al (2008) Repeated assessment of neuropsychological deficits in multiple sclerosis using the symbol digit modalities test and the MS neurospychological screening questionnaire. Mult Scler 14:940–946
18. Drake AS, Weinstock-Guttman B, Morrow SA, Hojnacki D, Munschauer FE, Benedict RH (2010) Psychometric and normative data for the multiple sclerosis functional composite: replacing the PASAT with the symbol digit modalities test. Mult Scler 16:228–237
19. Amato MP, Portaccio E, Goretti B et al (2010) Relevance of cognitive deterioration in early relapsing-remitting MS: a 3-year follow-up study. Mult Scler 16:1474–1483
20. Honarman K, Akbar N, Kou N, Feinstein A (2011) Predicting employment status in multiple sclerosis patients: the utility of the MS functional composite. J Neurol 258:244–249
21. Benedict RH, Cookfair D, Gavett R et al (2006) Validity of the minimal assessment of cognitive function in multiple sclerosis. J Int Neuropsych Soc 12:549–558
22. Rao SM, Leo GJ, Ellington L, Nauertz T, Bernardin L, Unveragt F (1991) Cognitive dysfunction in multiple sclerosis. II. Impact on employment and social functioning. Neurology 41:692–696
23. Swirsky-Sacchetti T, Mitchell DR, Seward J et al (1992) Neuropsychological and structural brain lesions in multiple sclerosis: a regional analysis. Neurology 42:1291–1295
24. Arnett PA, Rao SM, Bernardin L, Grafman J, Yetkin FZ, Lobeck L (1994) Relationship between frontal lobe lesions and Wisconsin Card Sorting Test performance in patients with multiple sclerosis. Neurology 44:420–425
25. Nocentini U, Rossini PM, Carlesimo GA et al (2001) Patterns of cognitive impairment in secondary progressive stable phase of multiple sclerosis: correlation with MRI findings. Eur Neurol 45:11–18
26. Edwards SG, Liu C, Blumhardt LD (2001) Cognitive correlates of supratentorial atrophy on MRI in multiple sclerosis. Acta Neurol Scand 104:214–223
27. Chiaravalloti ND, DeLuca J (2008) Cognitive impairment in multiple sclerosis. Lancet Neurol 7(12):1139–1151
28. Baddeley AD, Hitch GJ (1974) Working memory. In: Bower AG (ed) The psychology of learning and motivation: advances in research and theory. Academic, New York, pp 47–90
29. Rao SM, Aubin-Faubert P St, Leo GJ (1989) Information processing speed in patients with multiple sclerosis. J Clin Exp Neuropsyc 11:471–477
30. Demaree HA, De Luca J, Gaudino EA, Diamond BJ (1999) Speed of information processing as a key deficit in multiple sclerosis: implications for rehabilitation. J Neurol Neurosurg Psychiatry 67:661–663
31. Janculjak D, Mubrin Z, Brinar V, Spilich G (2002) Changes of attention and memory in a group of patients with multiple sclerosis. Clin Neurol Neurosurg 104:221–227
32. Penner IK, Raush M, Kappos L, Opwis K, Radu EW (2003) Analysis of impairment related functional architecture in MS patients during performance of different attention tasks. J Neurol 250:461–472
33. DeLuca J, Chelune GJ, Tulsky DS, Lengenfelder J, Chiaravalloti ND (2004) Is speed of processing or working memory the primary information processing deficit in multiple sclerosis? J Clin Exp Neuropsychol 26:550–562
34. Gronwall DM (1977) Paced auditory serial-addition task: a measure of recovery from concussion. Percept Mot Skills 44:367–373
35. Dujardin K, Donze AC, Hautecoeur P (1998) Attention impairment in recently diagnosed multiple sclerosis. Eur J Neurol 5:61–66
36. Baddeley AD (1990) Human Memory: theory and practice. Lawrence Erlbaum, London
37. Graf P, Schacter D (1985) Implicit and explicit memory for new associations in normal and amnesic patients. J Exp Psychol Learn Mem Cogn 11:501–518
38. Litvan I, Grafman J, Vendrell P et al (1988) Multiple memory deficit in patients with multiple sclerosis. Arch Neurol 45:607–610
39. Grigsby J, Ayarbe SD, Kravcisin N, Busenbark D (1994) Working memory impairment among persons with chronic progressive multiple sclerosis. J Neurol 241:125–131
40. Beatty WW, Monson N (1991) Metamemory in multiple sclerosis. J Clin Exp Neuropsychol 16:640–646

41. Beatty WW, Goodkin DE, Monson N, Beatty PA (1989) Cognitive disturbances in patients with relapsing remitting multiple sclerosis. Arch Neurol 46:1113–1119

42. Jennekens-Schinkel A, van der Velde EA, Sanders EA, Lanser JB (1990) Memory and learning in outpatients with quiescent multiple sclerosis. J Neurol Sci 95:311–325

43. Rao SM, Leo GJ, St Aubin-Faubert P (1989) On the nature of memory disturbance in multiple sclerosis. J Clin Exp Neuropsychol 11:699–712

44. De Luca J, Barbieri-Berger S, Johnson SK (1994) The nature of memory impairments in Multiple Sclerosis: acquisition versus retrieval. J Clin Exp Neuropsychol 16:183–189

45. De Luca J, Gaudino EA, Diamond BJ, Christodoulou C, Engel R (1998) Acquisition and storage deficits in multiple sclerosis. J Clin Exp Neuropsychol 20:376–390

46. Thornton AE, Raz N (1997) Memory impairment in multiple sclerosis: a quantitative review. Neuropsychology 11:357–366

47. Seinelä A, Hämäläinen P, Koivisto M, Ruutiainen J (2002) Conscious and unconscious uses of memory in multiple sclerosis. J Neurol Sci 198:79–85

48. Lezak M (1995) Neuropsychological assessment, 3rd edn. Oxford University Press, New York

49. Spreen O, Strauss E (1998) A compendium of neuropsychological tests. Administration, norms and commentary. Oxford University Press, New York

50. Pearson OA, Stewart KD, Aremberg D (1957) Impairment of abstracting ability in multiple sclerosis. J Nerv Ment Dis 125:221–225

51. Beatty PA, Gange JJ (1977) Neuropsychological aspects of multiple sclerosis. J Nerv Ment Dis 164:42–50

52. Heaton RK, Nelson LM, Thompson DS, Burks JS, Franklin GM (1985) Neuropsychological findings in relapsing-remitting and chronic-progressive multiple sclerosis. J Consult Clin Psychol 53:103–110

53. Rao SM, Hammeke TA, Speech TJ (1987) Wisconsin Card sorting test performance in relapsing-remitting and chronic-progressive multiple sclerosis. J Consult Clin Psychol 55:263–265

54. Birnboim S, Miller A (2004) Cognitive strategies application of multiple sclerosis patients. Mult Scler 10:67–73

55. Drew M, Tippett LJ, Starkey NJ, Isler RB (2008) Executive dysfunction and cognitive impairment in a large community-based sample with multiple sclerosis form New Zealand: a descriptive study. Arch Clin Neuropsychol 23:1–19

56. Foong J, Rozewicz L, Quaghebeur G et al (1997) Executive function in multiple sclerosis. The role of frontal lobe pathology. Brain 120:15–26

57. Vleugels L, Lafosse C, van Nunen A et al (2000) Visuoperceptual impairment in multiple sclerosis patients diagnosed with neuropsychological tasks. Mult Scler 6:241–254

58. Vleugels L, Lafosse C, van Nunen A, Charlier M, Ketelaer P, Vandenbussche E (2001) Visuoperceptual impairment in MS patients: nature and possible neural origins. Mult Scler 7:389–401

59. Kujala R, Portin R, Ruutiainen J (1996) Language function in incipient cognitive decline in multiple sclerosis. J Neurol Sci 141:79–86

60. Friend KB, Rabin BM, Groninger L, Deluty RH, Bever C, Grattan L (1999) Language functions in patients with multiple sclerosis. Clin Neuropsychol 1999(13):78–94

61. Friedman JH, Brem H, Mayeux R (1983) Global aphasia in multiple sclerosis. Ann Neurol 13:222–223

62. Achiron A, Ziv I, Djaldetti R, Goldberg H, Kuritzky A, Melamed E (1992) Aphasia in multiple sclerosis: clinical and radiologic correlations. Neurology 42:2195–2197

63. Arnett PA, Rao SM, Hussain M, Swanson SJ, Hammeke TA (1996) Conduction aphasia in multiple sclerosis: a case report with MRI findings. Neurology 47:576–578

64. Jonsdottir MK, Magnusson T, Kjartansson O (1998) Pure alexia and word-meaning deafness in a patient with multiple sclerosis. Arch Neurol 55:1473–1474

65. Grossman M, Robinson KM, Onishi K, Thompson H, Cohen J, D'Esposito M (1995) Sentence comprehension in multiple sclerosis. Acta Neurol Scand 92:324–331

66. Henry JD, Beatty WW (2006) Verbal fluency deficits in multiple sclerosis. Neuropsychologia 44:1166–1174

67. Ryan JJ, Prifitera A, Larsen J (1982) Reliability of the WAIS-R with a mixed patient sample. Percept Mot Skills 55:1277–1278
68. Ron MA, Callanan MM, Warrington EK (1991) Cognitive abnormalities in multiple sclerosis: a psychometric and MRI study. Psychol Med 21:59–68
69. Folstein MF, Folstein SE, McHugh PR (1975) "Mini mental state": a practical method for grading the cognitive state of patients for the clinician. J Psychiatr Res 12:189–198
70. Beatty WW, Goodkin DE (1990) Screening for cognitive impairment in multiple sclerosis. An evaluation of the mini-mental state examination. Arch Neurol 47:297–301
71. Beatty WW, Paul RH, Wilbanks SL, Hames KA, Blanco CR, Goodkin DE (1995) Identifying multiple sclerosis with mild or global cognitive impairment using the screening examination for cognitive impairment (SEFCI). Neurology 45:718–723
72. Rao SM (1990) Cognitive function study group. A manual for the brief repeatable battery of neuropsychological tests. National Multiple Sclerosis Society, New York
73. Solari A, Mancuso L, Motta A, Mendozzi L, Serrati C (2002) Comparison of two brief neuropsychological batteries in people with multiple sclerosis. Mult Scler 8:169–176
74. Randolph C (1998) Repeatable battery for the assessment of neuropsycholological status. Psychological Corporation, San Antonio
75. Aupperle RL, Beatty WW, Shelton F, deNap G, Gontovsky ST (2002) Three screening batteries to detect cognitive impairment in multiple sclerosis. Mult Scler 8:382–389
76. Carlesimo GA, Caltagirone C, Gainotti G, The Group for the Standardization of the Mental Deterioration Battery (1996) The mental deterioration battery: normative data, diagnostic reliability and qualitative analyses of cognitive impairment. Eur Neurol 36:378–384
77. Nelson HE (1976) A modified card sorting test sensitive to frontal lobe defects. Cortex 12:313–324
78. Smith A (2000) Symbol digit modalities test. manual. Webster Psychological Services, Los Angeles
79. Nocentini U, Giordano A, Di Vincenzo S, Panella M, Pasqualetti P (2006) The symbol digit modalities test – oral version: Italian normative data. Funct Neurol 21:93–96
80. Benedict RH, Fischer JS, Archibald CJ et al (2002) Minimal neuropsychological assessment of MS patients: a consensus approach. Clin Neuropsychol 16:381–397
81. Langdon D, Amato M, Boringa J, Brochet B, Foley F, Fredrikson S, Hämäläinen P, Hartung HP, Krupp L, Penner I, Reder A, Benedict R (2012) Recommendations for a brief international cognitive assessment for multiple sclerosis (BICAMS). Mult Scler 18:891–898
82. Salmon DP, Filoteo JV (2007) Neuropsychology of cortical versus subcortical dementia syndromes. Semin Neurol 27:7–21
83. Sumowski JF, Chiaravalloti N, Wylie G, Deluca J (2009) Cognitive reserve moderates the negative effect of brain atrophy on cognitive efficiency in multiple sclerosis. J Int Neuropsychol Soc 15:606–612
84. Sumowski JF, Wylie GR, Deluca J, Chiaravalloti N (2010) Intellectual enrichment is linked to cerebral efficiency in multiple sclerosis: functional magnetic resonance imaging evidence for cognitive reserve. Brain 133:362–374
85. Sumowski JF, Wylie GR, Chiaravalloti N, DeLuca J (2010) Intellectual enrichment lessens the effect of brain atrophy on learning and memory in multiple sclerosis. Neurology 74: 1942–1945
86. Moller A, Wiedemann G, Rohde U, Backmund H, Sonntag A (1994) Correlates of cognitive impairment and depressive mood disorder in multiple sclerosis. Acta Psychiatr Scand 89:117–121
87. Arnett PA, Higginson CI, Voss WD, Bender WI, Wurst JM, Tippin JM (1999) Depression in multiple sclerosis: relationship to working memory capacity. Neuropsychology 13:546–556
88. Arnett PA, Higginson CI, Randolph JJ (2001) Depression in multiple sclerosis: relationship to planning ability. J Int Neuropsychol Soc 7:665–674
89. Diamond BJ, Johnson SK, Kaufman M, Graves L (2008) Relationships between information processing, depression, fatigue and cognition in multiple sclerosis. Arch Clin Neuropsychol 23:189–199
90. Figved N, Benedict R, Klevan G et al (2008) Relationship of cognitive impairment to psychiatric symptoms in multiple sclerosis. Mult Scler 14:1084–1090

Conclusions

18

Ugo Nocentini, Gioacchino Tedeschi, and Carlo Caltagirone

The purpose of this volume is to provide state of the art knowledge about neuropsychiatric disorders in patients with Multiple Sclerosis. Because of the elusiveness and complexity of MS, before treating specific topics we have discussed some general aspects of the disease. Not only is a complete explanation of the etiology of the disease lacking, but many aspects of MS are still unclear. For example, there is no consistency in the correlations between characteristics of the lesions as they appear on MRI, and the symptomatic and functional picture presented by individual patients. Furthermore, the variety of damage to the nervous system caused by MS makes it difficult to construct a coherent set of information about the disease that would allow developing therapies and effective interventions to improve the lives of these patients. This is not surprising, however, because MS is a multiform disease of the nervous system, which is the most complex, and most difficult to evaluate, structure in the human body.

At the end of this book, therefore, we have more questions than answers. In any case, the clinical and experimental work that has been carried out on the neuropsychiatric disorders in MS has opened some small windows in the large edifice represented by the relationships between the structure of the nervous system and mental processes. Only the future will tell if we are able to pass from the limited perspective of observation offered by these windows to a broader vision.

U. Nocentini (✉) • C. Caltagirone
Dipartimento di Neuroscienze, Università degli Studi di Roma "Tor Vergata", Rome, Italy
e-mail: u.nocentini@hsantalucia.it; c.caltagirone@hsantalucia.it

G. Tedeschi
Istituto di Scienze Neurologiche, Seconda Università di Napoli, Naples, Italy
e-mail: gioacchino.tedeschi@unina2.it

U. Nocentini et al. (eds.), *Neuropsychiatric Dysfunction in Multiple Sclerosis*,
DOI 10.1007/978-88-470-2676-6_18, © Springer-Verlag Italia 2012

18.1   What Elements Do We Already Have at Our Disposal?

The following epidemiological data seem robust: Patients with MS present more disorders in the neuropsychiatric sphere than the general population and patients with equally invalidating illnesses. And the severity of these disorders is confirmed by the high suicide rate in patients with MS.

Depression is the neuropsychiatric disorder most studied in MS because of its high frequency and consequences. The present volume is no exception, due to the greater availability of data on depression.

These factors lead to a series of questions: Considering the epidemiological data and knowledge of the causes of several disorders, such as depression, what happens in MS? Is depression a reaction to the illness condition or is it directly caused by something (e.g. lesion localization or inflammatory processes) specific to MS? Depending on how we read the available data, we will favour one or the other hypothesis. Some studies have identified a prevalence of lesions in several cerebral structures in MS patients with depression or other psychic disorders compared to MS patients who do not present these disorders. The different frequencies of neuropsychiatric disorders between MS and other diseases should agree with these neurological data. But, considering the peculiar nature of MS, how can we discard the "reactive" hypothesis? This illness afflicts young, even very young, people who are still structuring their psychological reality. MS is not a "once and for all" disease and its history has an unpredictable plot and ending. It is thus not surprising that it provokes different reactions from those of other illnesses. The truth about the processes that lead to depression or to psychosis in MS might be found in a sort of middle way. Human beings interact with the experiences of life through the activity of specific cerebral structures. We know that brain networks exist which process specific emotional content from experiences, interacting with other networks that process different types of content. These experiences modify the activity of circuits that can also be modified by lesions (both micro- and macroscopic). A circuit altered by a lesion may "react" differently than a healthy circuit, and this can lead to differences in the result of processing an experience with respect to the normal situation. In MS, the elements exist that can lead to what we have outlined above. The classifying aspects of neuropsychiatric disorders are reported in the specific chapters. We would like to encourage researchers to go beyond diagnostic aspects and tackle the etiological interpretation of neuropsychiatric disorders. It might actually be more profitable to start from single signs and symptoms.

A recent study by Arnett et al. [1] is one of the first attempts to integrate the factors associated with depression in MS patients into a comprehensive model. The authors suggest that basic aspects of the disease are at the origin of depression in MS, including changes in cytology, histology, physiology and immunology of the CNS. In this regard, several different studies confirm that inflammatory and oxidative processes have an important role in the genesis of primary depressive disorder [2]. Among the consequences of these basic factors, apart from depression itself, they considered fatigue, cognitive dysfunctions, physical disability and pain. There are significant, also bi-directional, interactions among these elements. As the significance of these interactions was below expectations, Arnett et al. hypothesized that they are moderated by other factors. Reports in the literature suggest the following as moderators: social support, coping, stress, self-concept and illness-concept. The moderating action can be carried out by the individual moderator or by interactions among moderators. It is preferable to speak of moderators rather than mediators, because factors such as stress and social support do not seem to have a causal (mediating) role in depression, but lead to an increase or a decrease in the effects of other factors.

No empirical data support all the aspects of this model, but many of its assumptions could be tested experimentally in the future. In any case, the model is a good example of the complexity of the factors involved in the relationship between depression and MS. On the one hand, the data reported in various parts of this volume show that accuracy and specificity in gathering clinical data are increasing and, on the other, that technology is providing increasingly sophisticated instruments and methods of investigation. The union of these two elements should bring us closer to understanding the causes of the phenomena and the possibility of identifying effective solutions. Indeed, we are close to reaching the goal of individualized therapy.

Nevertheless, other data bring us back to the current reality and the need to work as best we can with the instruments available. Several years ago both a review in the literature and a Consensus Conference [3,4] pointed out how depression was underestimated in patients with MS and the consequences of this at the therapeutic level. We feel this should be brought back to the forefront of attention, because of the harmful consequences of continuing to underestimate the phenomenon.

Although we have no specific or adequate proof of the effectiveness and risk of the various options for treating neuropsychiatric disorders in patients with MS, we cannot use this as a reason to neglect doing what we are already able to do. At the same time (and we hope this book will make a contribution in this direction), we should try to promote the appropriate trials to evaluate therapeutic options.

There are people who will certainly be grateful for our commitment.

References

1. Arnett PA, Barwick FH, Beeney JE (2008) Depression in multiple sclerosis: review and theoretical proposal. J Int Neuropsychol Soc 14:691–724
2. Maes M, Kubera M, Obuchowicswa E, Goelher L, Brzeszcz J (2011) Depression's multiple comorbidities explained by (neuro)inflammatory and oxidative & nitrosative stress pathways. Neuroendocrinol Lett 32(1):7–24
3. Siegert RJ, Abernethy DA (2005) Depression in multiple sclerosis: a review. J Neurol Neurosurg Psychiatry 76:469–475
4. Goldman Consensus Group (2005) The Goldman Consensus statement on depression in multiple sclerosis. Mult Scler 11:328–337

Index

A

ACTH, 102
Active plaques, 27
Acute disseminated encephalomyelitis
 (ADEM), 33
ADC. *See* Apparent diffusion coefficient
 (ADC)
Adverse effects, 114–117
Affective disorders, 122, 123
Age of onset, 9
Alternating attention, 135
Amantadine, 71
Amyloid precursor protein, 29
Anger, 127–129
Annual social cost, 3
Anxiety, 85–94, 108
Apparent diffusion coefficient (ADC), 47, 48
Atrophy, 46, 49, 50, 52–54, 108, 109
Attention, 133–137, 139, 141, 144–146, 149
Autonomic dysfunction(s), 16
Axonal damage, 29
Axonal damage/loss/injury/transection, 44, 45,
 47, 49–52
Azathioprine, 68

B

Baclofen and tizanidine, 70
BBB. *See* Blood-brain barrier (BBB)
BDI. *See* Beck Depression Inventory (BDI)
Beck Depression Inventory (BDI), 92, 108
Benign MS, 12
Bipolar disorder (BD), 99–104
 type I (BD I), 100
 type II (BD II), 100
Black holes (BHs), 44–46, 50, 108
Bladder dysfunction(s), 12, 16
Blood-brain barrier (BBB), 44

Blood oxygenation level-dependent (BOLD),
 49
Borderline personality disorder, 103
Botulinum toxin, 70, 73
Bowel dysfunction, 16
Brainstem dysfunction(s), 15
Brief repeatable battery of neuropsychological
 tests (BRBNT), 146

C

Cambridge Basic MS Score (CAMBS), 39
Carbamazepine, 71, 72
Carbolithium, 104
CBT. *See* Cognitive behavior therapy (CBT)
CCSVI. *See* Chronic cerebrospinal venous
 insufficiency (CCSVI)
CD4 and CD8 T cells, 23
Central executive system, 137
Cerebellar dysfunction(s), 14, 15
Cerebrospinal fluid (CSF), 44, 49, 50, 109
cGM-FLs. *See* Cortical gray matter focal
 lesions (cGM-FLs)
Charcot, J.-M., 4
Charcot's triad, 4
Choline, 48
Chronic active plaques, 27
Chronic cerebrospinal venous insufficiency
 (CCSVI), 23, 24, 70
CIS. *See* Clinically isolated syndrome (CIS)
Clinical course, 113, 115, 116
Clinically isolated syndrome (CIS), 12, 32,
 33, 108
Clusters, 9
Cognitive behavior therapy (CBT), 94
Cognitive disability, 49, 52–54
Cognitive impairment (CI), 54, 55, 90, 91
Cognitive rehabilitation, 77, 79–80

Co-morbidity, 99, 101
Comprehensive model (of depression in MS),
    156
Computed tomography (CT), 107
Conventional MRI (C-MRI), 43–46
Cortical GM (cGM), 49, 50, 52–54
Cortical gray matter focal lesions (cGM-FLs),
    50, 52
Cortical-juxtacortical lesions, 28
Cortical thickness (CTh), 49, 54
Corticosteroids, 101, 102, 104
Corticotropin-releasing hormone (CRH), 108
Creatine (Cr), 48, 49
CRH. See Corticotropin-releasing hormone
    (CRH)
Cruveilhier, J., 4
Cyclophosphamide, 68–69
Cytotoxic and suppression T lymphocytes, 23

D
Deep gray matter (dGM), 53, 54
Delusions, 115–117
Demyelinating disease(s), 3, 4
Demyelinating plaque, 27
Demyelination, 51
Depression (T2, T1, T2/T1-lesion load),
    107–109
Desipramine, 93
Devic's optic neuromyelitis, 34, 35
    criteria, 34
Diagnosis, 31–35
Diagnostic criteria, 31, 32, 34, 43, 44, 46
Diagnostic hypothesis, 115, 117
Diaschisis,
Differential diagnosis, 34, 35, 116
Diffusion tensor imaging (DTI), 47–48, 51–53,
    109
DIR. See Double inversion recovery (DIR)
Disease modifying therapy, 65–70
Divided attention, 135, 136
Double inversion recovery (DIR), 50
DR15 and DQ6 haplotypes, 21
DR4 haplotype, 22
DSM-IV, 109
DSM-IV-TR, 86, 100–102
DTI. See Diffusion tensor imaging (DTI)

E
EAE. See Experimental autoimmune
    encephalomyelitis (EAE)
EBV. See Epstein-Barr virus (EBV)

EDSS. See Expanded Disability Status Scale
    (EDSS)
Emotional lability, 103, 122, 123
Encoding, 136, 139, 140
Epstein-Barr virus (EBV), 22
Euphoria, 121–124
Executive function(s), 128, 129, 135, 138, 141,
    142, 144, 147, 149
Expanded Disability Status Scale (EDSS),
    38–40, 45, 46
Experimental autoimmune encephalomyelitis
    (EAE), 21–23

F
FA. See Fractional anisotropy (FA)
Fampridine, 71
Fatigue, 16, 17, 54, 86, 87, 89–90, 92, 94
Fatigue Severity Scale (FSS), 40
Fingolimod, 69
fMRI. See Functional MRI (fMRI)
Focal WM lesions, 43–45, 47, 52, 54
Focused attention, 135
Follicle-like structures, 28
Fractional anisotropy (FA), 48, 51–53, 109
Freesurfer, 50
Frontal lobe, 108
FSS. See Fatigue Severity Scale (FSS)
Functional MRI (fMRI), 43, 49, 54–55, 109,
    110

G
Gabapentin and pregabalin, 71, 72
Gadolinium (Gd), 44–46
General intelligence, 144
Generalized anxiety disorder, 86, 87
Genetic factors, 21–22
Geographic distribution, 8
Glatiramer acetate (GA), 67, 71
Gliosis, 27
Global GM, 49, 50, 53–55
Gray matter (GM), 43, 47–50, 52–55, 109

H
Hallucinations, 116, 117
Hamilton Depression Rating Scale (HDR-S),
    92
Hamilton Rating Scale for Depression
    (HAM-D), 107, 108
Headache, 17
Healthy controls (HCs), 51–55, 108

HHV6. *See* Human herpes virus 6 (HHV6)
High-field MRI, 49
High-risk areas, 8
HLA system, 102
1H-MRS. *See* Proton magnetic resonance
    spectroscopy (1H-MRS)
Human herpes virus 6 (HHV6), 22
Hypothalamic-pituitary-adrenal axis (HPA),
    108

I
Illness Severity Scale, 39
Immunological factors, 22–23
Implicit memory, 137, 138, 140
Inactive plaques, 27
Incapacity Status Scale, 39
Incidence, 7–9, 85, 86
Inflammatory cytokines, 91
Information processing speed, 133, 135, 136,
    138–140, 146, 147, 149
Integration, 79
Interdisciplinary teams, 79
Interferon, 88–89, 91
β-Interferon, 66–67
Inter-nuclear
    ophtalmoplegia, 15
Irritability, 87, 88, 92, 100, 103

K
Kurtzke Disability Status Scale (DSS), 38

L
Lactate, 49
Language, 135, 143–147
Lhermitte's sign, 14
Linkage studies, 22
Lipid-laden macrophages, 27
Long-term memory (LTM), 136, 137, 139,
    145, 146
Low-risk areas, 8
Lymphoid follicle-like structures, 22

M
MACFIMS, 147
Macrophage infiltrates, 27
Magnetic resonance imaging (MRI), 31–35,
    43–52, 54, 107–110

Magnetization transfer imaging (MTI), 47, 48,
    50–53
Magnetization transfer ratio (MTR), 47, 50–53
Major depressive disorder (MDD), 85, 87,
    93, 94
Malignant MS, 12
Mania, 99–104, 121
Mania Rating Scale (MRS), 103
MDD. *See* Major depressive disorder (MDD)
Mean diffusivity (MD), 48, 51–53, 109
Medium-risk areas, 8
Mental deterioration battery, 146–147
Meta-memory, 138
Methotrexate, 69
Methylprednisolone (MP), 65
MFIS. *See* Modified Fatigue Impact Scale
    (MFIS)
MHC alleles, 21
Migration, 9
Mini mental state examination (MMSE), 145
Mitoxantrone, 68
MMSE. *See* Mini mental state examination
    (MMSE)
Moderators, 157
Modified Fatigue Impact Scale (MFIS), 40
Monoclonal antibodies, 67–68
Mood dysfunction (MD), 107–110
Mortality rates, 8
MPRAGE, 50
MRI. *See* Magnetic resonance imaging (MRI)
MS. *See* Multiple sclerosis (MS)
MSIS-29. *See* Multiple Sclerosis Impact Scale
    (MSIS-29)
MSQoL-54. *See* Multiple sclerosis quality of
    life (MSQoL)-54
MTI. *See* Magnetization transfer imaging
    (MTI)
MTR. *See* Magnetization transfer ratio (MTR)
Multidisciplinary, 77, 79
Multiple sclerosis (MS), 3, 4, 43–55, 107–110
    subtypes, 11–12
Multiple sclerosis functional composite
    (MSFC), 39, 40, 134
Multiple Sclerosis Impact Scale (MSIS-29), 40
Multiple sclerosis quality of life (MSQoL)-54,
    40
Myelin basic protein (MBP), 23
Myelin debris, 27, 28
Myelin oligodendrocyte glycoprotein (MOG),
    23
Myo-inositol, 49, 51, 52

N

NAA. *See* N-acetylaspartate (NAA)
N-acetylaspartate (NAA), 48–53
Natalizumab, 67, 68
Negative symptoms, 116
Neuroimaging, 113, 114, 117–118
Neuroinflammation, 91
Neurological Rating Scale (SCRIPPS), 39
Neuroplasticity, 43, 45, 49, 55
Neuropsychological screening battery for
    multiple sclerosis (NPSBMS), 146
Neuropsychological situation, 90
Neutralizing antibodies, 66
Nine-hole peg test (9HPT), 39, 40
Non-conventional MRI (nc-MRI), 43, 47–55
Normal-appearing brain tissue (NABT), 109
Normal appearing gray matter (NAGM), 109
Normal-appearing white matter (NAWM), 43,
    47, 51–52, 109
NPSBMS. *See* Neuropsychological screening
    battery for multiple sclerosis
    (NPSBMS)

O

Olanzapine, 104
Onset symptom, 114–116
Optic neuritis, 12–14, 17
Oral therapies, 69
Organizational modalities, 78

P

Pain, 14, 15, 18
Panic disorder, 86, 87
Paroxysmal symptoms, 12, 13, 17–18
Pathological laughing and crying, 121–124
PD images/sequences, 44
Perivascular lymphocytes, 27
PET, 108
Physical disability, 49, 52–54
Physical exercise, 94
PML. *See* Progressive multifocal
    leukoencephalopathy (PML)
Positive, 115
Prefrontal cortex (PFC), 101, 109
Prevalence, 7–9, 85, 86, 88, 89, 91
Primary progressive, 12
Prognosis, 34, 35
Progressive multifocal leukoencephalopathy
    (PML), 67, 68
Progressive-relapsing, 12
Proton magnetic resonance spectroscopy
    (1H-MRS), 48–53

Pseudo-bulbar syndrome, 15
Psychiatric relapses, 113
Psychiatric symptoms, 113–117
Psychodiagnostic evaluation, 116
Psychotic disorders, 114, 115
Pyramidal dysfunctions, 13, 16

R

Ratio, male/female, 9
RBANS. *See* Repeatable Battery for the
    Assessment of Neuropsychological
    Status (RBANS)
Reactive hypothesis, 156
Rehabilitation, 77–80
Relapse, 32, 35
Relapse therapy, 65
Relapsing-remitting MS (RRMS), 11–12
Reliability, 37, 40
Remyelination, 28, 29
Repeatable Battery for the Assessment of
    Neuropsychological Status (RBANS), 146
Reproducibility, 37, 39
Retrieval, 137, 139

S

Saint Ludwina, 4
Screening examination for cognitive
    impairment (SEFCI), 145, 146
Secondary progressive, 12
SEFCI. *See* Screening examination for
    cognitive impairment (SEFCI)
Seizures, 18
Selective attention, 135, 137
Sensory dysfunction(s), 13, 14
Sertraline, 93
Sexual dysfunction, 13, 16
Shadow plaques, 28
Short tau inversion recovery (STIR), 44
Short-term memory, 137, 138, 145, 146
SIENA. *See* Structural image evaluation using
    normalization of atrophy (SIENA)
SIENAX, 50
Sleep disorders, 18
Social phobia, 86, 87
Spasticity, 13
SPECT, 108, 109
Spinal cord injuries (SCI), 107
SPM. *See* Statistical parametric mapping
    (SPM)
State-Trait Anger Expression Inventory
    (STAXI), 128
Statistical parametric mapping (SPM), 50

STAXI. *See* State-Trait Anger Expression
    Inventory (STAXI)
Stem completion, 138, 140
STIR. *See* Short tau inversion recovery (STIR)
Storage, 139
Structural image evaluation using
    normalization of atrophy (SIENA), 50
Sub-cortical cognitive impairment, 145
Subpial lesions, 28
Suicide ideation, 89
Sustained attention, 135–137, 146
Symptomatic therapy, 65, 70–73

T
T and B lymphocytes, 22
Temporal lobe, 107, 108
T2-FLAIR images/sequences, 44
Theory of mind, 129
3 s-Paced auditory serial addition test
    (PASAT 3), 39
T1 images/sequences, 44–46, 50, 51
T2-images/sequences, 44–46, 50
Timed 25-ft (T-25f) walk, 39
Tractography, 48
Transitional MS, 12

Treatment, 114, 115, 117
Troiano Functional Scale, 39
2D/3D FLAIR, 50

U
Uhtoff's phenomenon, 15

V
VBM. *See* Voxel-based morphometry (VBM)
Viral factors, 22
Visual dysfunction, 14–15
Visual memory, 139, 140
7/24 Visual Memory Test, 139
Visuo-spatial functions, 142–143
Voxel-based morphometry (VBM), 50, 54

W
WBNAA. *See* Whole brain N-acetylaspartate
    (WBNAA)
White matter (WM), 43–45, 47–54,
    108, 109
Whole brain N-acetylaspartate (WBNAA), 53
WM. *See* White matter (WM)
Working memory, 134–138, 146, 149